THE LIVES OF HAWAI'I'S DOLPHINS AND WHALES

THE LIVES OF HAWAI'I'S DOLPHINS AND WHALES

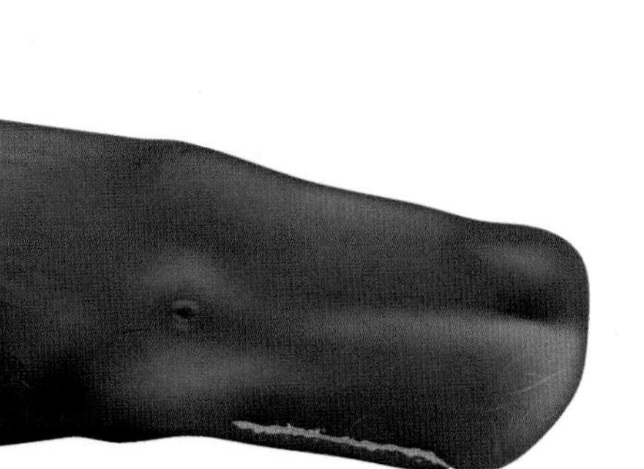

NATURAL HISTORY AND CONSERVATION

Robin W. Baird

UNIVERSITY OF HAWAI'I PRESS
HONOLULU

Printed in China

21 20 19 18 17 16 6 5 4 3 2 1

Library of Congress Cataloging-in-Publication Data

Names: Baird, Robin W., author.
Title: The lives of Hawaiʻi's dolphins and whales : natural history and conservation / Robin W. Baird.
Description: Honolulu : University of Hawaiʻi Press, [2016] |
| Includes bibliographical references and index.
Identifiers: LCCN 2016017725 | ISBN 9780824859985 (pbk. ; alk. paper)
Subjects: LCSH: Dolphins–Hawaii. | Whales–Hawaii.
Classification: LCC QL737.C43 B35 2016 | DDC 599.509969–dc23
LC record available at https://lccn.loc.gov/2016017725

University of Hawaiʻi Press books are printed on acid-free paper and meet the guidelines for permanence and durability of the Council on Library Resources.

ILLUSTRATIONS BY UKO GORTER
DESIGNED BY JULIE MATSUO-CHUN

CONTENTS

CONTENTS

PREFACE

Nai'a and *koholā*—the Hawaiian names for dolphins and whales, respectively—have always been well known to Native Hawaiians. Dolphins and whales appear in ancient chants and songs, evidence that they were respected in Hawaiian culture. The humpback whale, *koholā kuapi'o,* was revered as a manifestation of Kanaloa, the Hawaiian god of the oceans. The first *nai'a* was a child of Kanaloa and Papa, the earth mother. Although I am not Hawaiian and live in the islands only part time, over the last seventeen years I've had the opportunity to spend much of my time working with dolphins and whales in Hawaiian waters, the last thirteen years while working with Cascadia Research Collective, a nonprofit organization based in Olympia, Washington. This work has typically involved spending hundreds of hours each year on the water looking for, finding, and working with various species, recording behavior, photographing individuals, collecting biopsy samples for genetic and toxicology studies, attaching suction-cup tags to study short-term behavior, and, since 2006, satellite tagging individuals to examine longer-term behavior and movements. But for every hour spent on the water, substantially more time is spent working with the data collected in the field: hundreds of thousands of photographs, satellite tag data from over two hundred individuals, and dive data from almost a hundred individuals.

When I started working in Hawai'i in 1999, there was a crowded research environment, but with one or two exceptions, every publication on whales and

On August 21, 2014, two striped dolphins, an open-ocean species, were seen in shallow water off Kihei, Maui. One of the dolphins stranded and died and the other was not seen again. This photo shows Kimokeo Kapahulehua, a Native Hawaiian cultural practitioner, saying a prayer over the dolphin that died. Photo by Ed Lyman/NOAA.

dolphins in Hawaiian waters in the previous thirty years had focused on just two species: humpback whales or spinner dolphins. With an incredible diversity of poorly known species close at hand, it made no sense for me, a newcomer to the islands, to focus on those well-studied species, and when opportunities arose to work with any of the rarer, lesser-known species, I took them. Although six species of baleen whales, other than humpbacks, have been recorded in Hawaiian waters, all six are rarely encountered, and my work thus focused almost entirely on toothed whales and dolphins—the odontocetes.

Today we know an incredible amount about the majority of the species of odontocetes in Hawaiian waters based both on our work over the last seventeen years and the work of others. Our research has been a team effort—the biopsy samples we've collected, almost 1,500 samples from

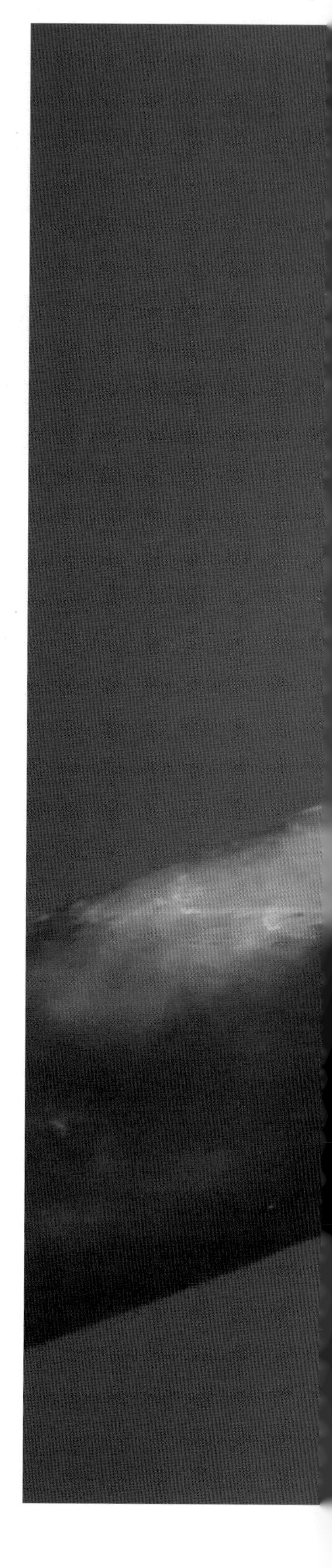

Short-finned pilot whale mother and newborn calf off Kona, Hawai'i Island, May 17, 2009. The mother, HIGm0054 in our photo-identification catalog, was first identified off the island in May 2003. Although HIGm0054 was seen several times in 2010, the calf was not present and probably did not survive. Photo by Deron S. Verbeck/iamaquatic.com.

fourteen different species, have gone to geneticists, toxicologists, and researchers studying stable isotopes and the microbiome, among other things. Graduate students and others have worked with the hundreds of thousands of photographs, establishing catalogs of distinctive individuals of twelve different species and using photos to examine scarring patterns, estimating age of individuals, examining social organization, and producing abundance estimates. These catalogs are curated through Cascadia Research, but when I refer to our research, I include both Cascadia's studies and those of our collaborators.

This book is an attempt to summarize much of what is known about dolphins and whales in Hawaiian waters, making it more easily available for the general public and those interested in Hawaiian natural history. It is, admittedly, biased toward my own work through Cascadia, or work I've been involved with through collaborations with other researchers, but I have tried to bring in important findings from other researchers and observations from others who have spent time on the water in Hawai'i. It is also biased toward the main Hawaiian Islands, the eight large islands in the eastern part of the archipelago. Comparatively little research has been undertaken on whales or dolphins around the Northwestern Hawaiian Islands, with the exception of several studies of spinner dolphins and large-vessel surveys undertaken by the National Marine Fisheries Service, and sadly I've not had the opportunity to work there.

My approach to working with Hawaiian odontocetes has been quite different from most other researchers studying whales or dolphins worldwide. Rather than focusing on the most easily studied species or trying to address one particular question—a good strategy if you are a graduate student—when we've encountered rare species, we've always tried to make the most of the opportunity, staying with the rarest species the longest. As a short-term strategy this would have been a bad idea, as obtaining enough information on rare species to say something conclusive about them obviously would take a long time. But while my work in Hawai'i started as a short-term job—in part to get away from another winter in Halifax, Canada—it turned into a long-term study. Thus taking advantage of those rare opportunities in the early years has turned out to be a great long-term approach to learn about some otherwise poorly known species: rough-toothed dolphins, dwarf sperm whales, false killer whales, Cuvier's beaked whales, pygmy killer whales, melon-headed whales, and others.

OASIS IN A DESERT SEA

The Hawaiian archipelago is the most isolated island group in the world, and unlike many islands today it has always been isolated, formed by a volcanic hot spot in the middle of the Pacific Ocean, rather than by cleaving off a portion of a larger land mass. The newest islands, the eight main Hawaiian Islands, lie within the tropics in the southeastern part of the chain, while most of the Northwestern Hawaiian Islands, extending as far as 2,200 kilometers (km) (1,368 miles [mi]) to the northwest of Honolulu on the island of O'ahu, lie above the Tropic of Cancer. The isolation of the islands and their independent origin resulted in few colonizers, but their remoteness has also led to many examples of endemism—development of unique and specialized forms of life in the absence of other competitors or predators. The islands are particularly famous for their populations of endemic birds and plants, and the distance of the islands from continental landmasses effectively prohibited their colonization by terrestrial mammals, reptiles, or amphibians—with the exception of one species of bat—until the arrival of the Polynesians.

Surrounding the islands is an immense expanse of unproductive and deep ocean waters. In the marine environment, the distance from other shallow-water areas has limited the diversity of fish, coral, and invertebrate species found around the Hawaiian Islands. In terms of cetaceans—the whales (*koholā* or *palaoa* in Hawaiian) and dolphins (*nai'a* in Hawaiian)—diversity in tropical waters is higher than in more temperate areas, and for long-lived, fast-moving species

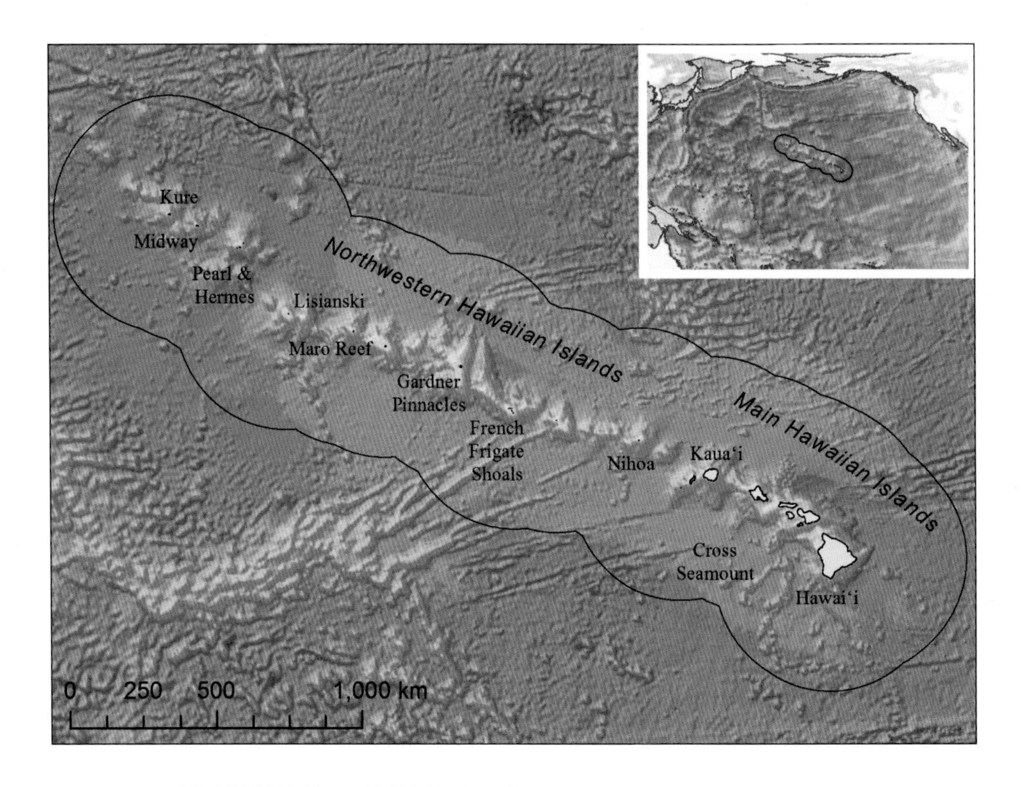

The Hawaiian Islands extend over 2,500 kilometers (1,550 miles) from Kure Atoll to Hawai'i Island. The boundary shown is the 200-nautical-mile U.S. Exclusive Economic Zone (EEZ) surrounding the islands.

dispersal is not so limited by deep waters or distance. Whales and dolphins were probably the first visitors to the islands, soon after they formed, and among the first to make the island slopes their home. The steep slopes of the islands, resulting from their volcanic origin in the middle of the Pacific plate, means that the deepwater home of many species of open-ocean whales and dolphins is actually very close to shore in Hawai'i. Such proximity should make for a good opportunity to study otherwise open-ocean species of whales and dolphins, but the productivity of the central tropical Pacific is so low that the density of individuals of different species of whales and dolphins around the islands is incredibly low. Despite this, there are many resident populations of whales and dolphins around the Hawaiian archipelago.

THE OASIS

Several thousand years ago, Pacific Islanders in their voyaging canoes spread out among the islands of the South and central Pacific. Crossing vast expanses of water, these superb navigators used the cues of their environment to find land—ocean swells, cloud patterns, and seabirds. Approaching the Hawaiian Archipelago by sea, Polynesian voyagers, sailors, and oceanographers have all crossed this enormous expanse of open water that has relatively little life. The waters of the central tropical Pacific are oligotrophic, meaning that they contain few nutrients. A glance at a map of the North Pacific reveals a vast blue area of open ocean, in the middle of which lie the Hawaiian Islands. The map does not reveal, however, that there is very little else in the water. The reason that waters are so clear in Hawaiʻi is that they hold less phytoplankton and zooplankton, the basis of the oceanic food web. Not surprisingly, biomass and productivity in the open ocean are a small fraction of that in tropical rain forests. But, in fact, productivity in the central tropical Pacific is also only a small fraction of that in the tundra or even in terrestrial deserts. Compared to coastal or higher latitude marine areas, essentially much of the central tropical Pacific is a biological desert.

The Hawaiian Archipelago was formed by a hot spot in the earth's mantle, an area where magma close to the surface resulted in a series of volcanic eruptions that were sustained long enough to form the islands. The hot spot remains stationary, but with the movements of the earth's crust over millions of years, an arc of islands was formed stretching over thousands of kilometers. The size and height of each island was determined by the intensity and duration of the volcanic eruptions. Today, after millions of years of erosion and settling, the islands to the northwest have been reduced to seamounts or coral atolls, and only a handful of the more recent islands project far above the water's surface. These islands, alone in the middle of the Pacific, create their own "weather," both above and below the water's surface.

Map showing chlorophyll-a levels in the central tropical Pacific surrounding the Hawaiian Islands, indicating slightly higher productivity around some of the islands and in the subtropical convergence zone to the north. This map was produced using data from 2003 through 2014 in the first quarter of each year. Courtesy of Pacific Islands Fisheries Science Center.

Currents in the ocean are complex, driven at the surface by wind and at depth by differences in water density (driven itself by differences in temperature). Around the main Hawaiian Islands, the eight large islands at the southeastern end of the archipelago, the trade winds blow predominantly from the northeast, and surface currents flow from east to west. Deeper in the water column, currents driven by differences in water density also flow from east to west. The islands, rising steeply some 4,000 meters (m) (13,000 feet) from the ocean floor, disrupt these currents both at depth and at the surface. Meeting the islands, deep ocean currents rise toward the surface, bringing cooler water with higher levels of nutrients. Winds over the islands are deflected and influenced by the mountain masses. For example, wind speeds may double off the northern and southern ends of the island of Hawaiʻi. As moisture-laden clouds reach the islands, they rise and release their moisture in the form of rain, which washes nutrients into the ocean. These upwelling currents, strengthened winds that create eddies, and increased nutrients from runoff combine to create the oasis.

The waters of the open ocean are strongly stratified—surface waters are mixed by winds but are warm and oxygen poor. While sunlight penetrates near-surface waters, these waters are low in nutrients, and phytoplankton populations, the basis of the food web, are small. Organisms die and sink below the photic zone, settling in deep, dark, cold waters below. Productivity around the main Hawaiian Islands is still comparatively low, with only a slight increase in surface chlorophyll detected. Yet if they are surrounded by a biological desert, even an area that has low productivity will be an oasis to many organisms. The stronger winds funneling around the islands create a complex field of eddies, some of which drive upwelling of the colder nutrient-rich waters to the surface, where sunlight allows phytoplankton to bloom. The islands themselves create an "island mass" effect, deflecting the cold, nutrient-rich, deeper waters upward. We typically think of oases within a desert on land resulting from the presence of water, allowing the growth and persistence of life where not possible elsewhere. It is not the productivity itself around the islands that creates the oasis; it is the discontinuity between the extremely low productivity waters of the central tropical Pacific and the slightly more productive waters immediately surrounding the islands that creates it.

Of the twenty-five species of whales and dolphins that have been recorded in Hawaiian waters, more than half (eighteen) of the species are odontocetes—the toothed whales and dolphins. The remaining seven species are mysticetes—the baleen whales. Most of the baleen whales come to Hawaiian waters only in the winter, and some of the odontocetes just move through the area, part of large open-ocean populations. But for eleven of the species, all of them odontocetes, this oasis has created a year-round home, with populations living off the island shores, taking advantage of the increased predictability of prey.

HOW WE STUDY WHALES AND DOLPHINS

What we know about dolphins and whales in Hawai'i comes largely from how they have been studied—the methods or techniques we use to gain information about their lives or populations. Surveys in offshore waters are undertaken with large vessels (typically a couple of hundred feet long). Throughout the species accounts that follow, I refer to results from two large vessel surveys undertaken by the National Marine Fisheries Service (NMFS), one in 2002 and one in 2010. Both were of five months' duration, and each ship was at sea for up to a month between port calls. The surveys involved a team of observers using very large, twenty-five-power binoculars (called "Big-Eyes"), scanning the water and recording all sightings. These are called line transect surveys, and each vessel travels along predetermined transect lines. Using the group sizes, the sea conditions, how far away from the vessel a group is first spotted, and a lot of complicated math, the abundance of most species was estimated. These estimates are the primary abundance estimates for almost every species of whale or dolphin in Hawaiian waters. They are required under the Marine Mammal Protection Act in order to determine the status of each species.

Both the 2002 and 2010 surveys were undertaken in the summer and fall, August through November. Although they do a good job of sampling species that are in Hawaiian waters year-round, such surveys miss a lot of the migrating baleen whales, as these whales use Hawaiian waters primarily in the winter. In addition, conditions in offshore Hawaiian waters are typically very windy, so sightings of difficult-to-spot species, such as beaked whales, dwarf sperm whales, or pygmy killer whales, are infrequent, and thus some of the estimates have a lot more uncertainty associated with them than others. The two surveys both covered all of the Hawaiian waters, from the islands out to 200 nautical miles from shore,

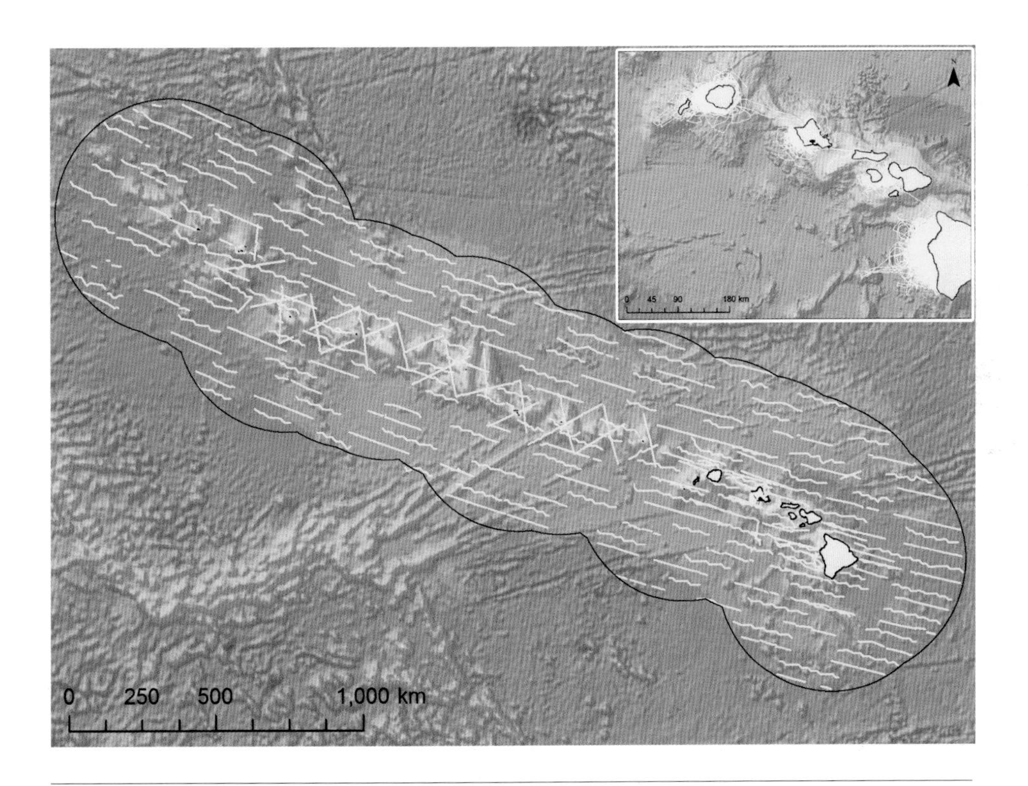

Almost all the survey efforts in offshore Hawaiian waters are from National Marine Fisheries Service (NMFS) large-vessel surveys, while small-boat-based surveys have primarily been undertaken around the main Hawaiian Islands. This map shows tracklines from all NMFS whale and dolphin surveys since 1986, with tracklines from our small-boat efforts from 2000 through mid-2015 shown in the inset.

covering all of the U.S. Exclusive Economic Zone (EEZ). I refer to this larger area, covering 2.45 million square kilometers (km^2), as "all Hawaiian waters," but it is important to realize that this is a huge area, almost equivalent in size to the combined U.S. EEZs off the lower forty-eight states—together just 2.52 million km^2.

When the conditions are good and a high-priority species is spotted, small boats are often launched from the larger survey vessel—biologists on board take photos and collect small biopsy samples. These samples are collected remotely, typically using a crossbow and a biopsy arrow (or dart) that includes a small hollow tip on the end. The dart tip penetrates the skin and blubber and bounces off, taking a piece of skin the size of a pencil eraser and usually a small amount of blubber. The dart, hopefully with a sample, is then picked up out of the water and the sample is stored for later analysis. When the dart strikes, the whale or dolphin typically flinches and may dive faster, but the effects seem to be short lived, and the biopsy sites heal quickly. This work, like the surveys themselves and the tagging work discussed below, is all performed under research permits issued by the NMFS Office of Protected Resources that sanction the use of these techniques. The skin samples are usually used for genetic analyses to examine differences among populations, determine the sex of individuals, and for some species, to look at more details of their social organization, such as mating habits and relationships between individuals within groups. The blubber samples are usually used to look at pollutant levels, but they can also be used to look at hormones. For example, it's possible to determine whether a female is pregnant or a male is sexually mature, as well as how stressed the individual is.

These large vessels also usually tow a series of hydrophones—underwater microphones—that monitor the sounds that different species make. Many species make sounds that are unique to them, such as the echolocation clicks of beaked whales or the "boings" of minke whales. Autonomous acoustic recorders can be used to characterize species and allow for monitoring. They can be deployed on ocean gliders or on instruments mounted on the seafloor and later recovered. One of these seafloor acoustic recorders, a High-frequency Acoustic Recording Package (HARP), has been deployed off Kona, the west side of Hawai'i Island, almost continuously since 2006.

In our work around the main Hawaiian Islands and in most of the nearshore studies, we use smaller boats. Over the years we've used boats ranging in size

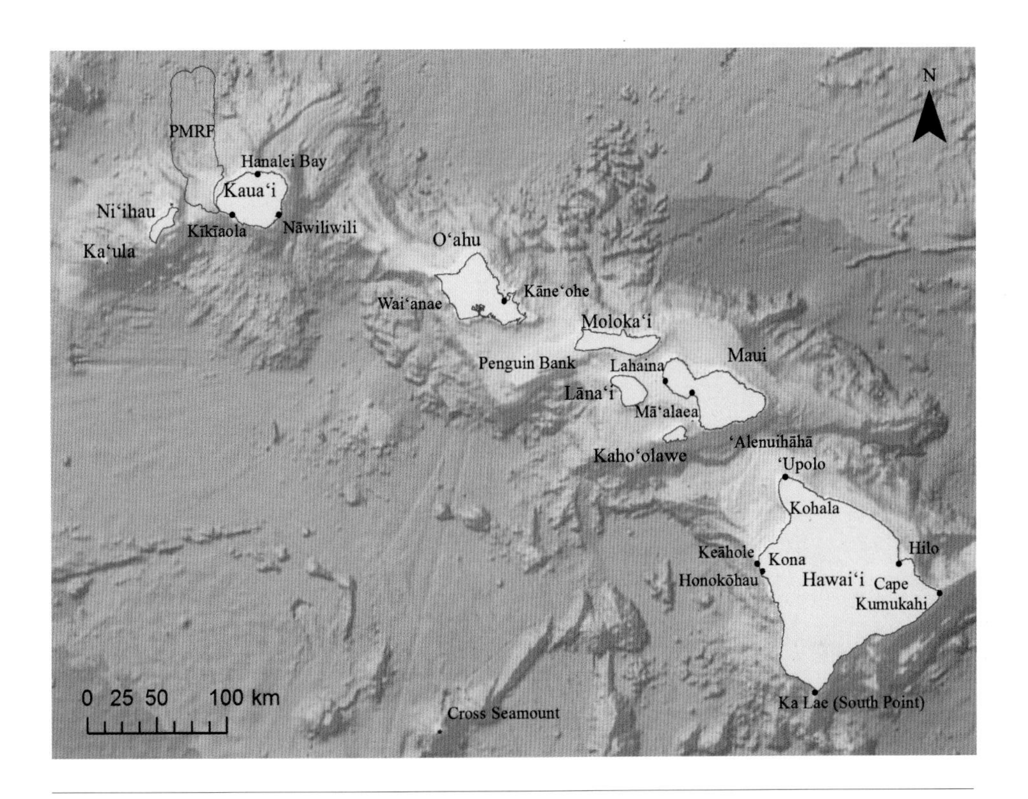

The main Hawaiian Islands, showing place names mentioned in the text.

from 11′ to 60′, but our primary research vessels are an 8.2-m (27′) Boston Whaler that we use mainly off Kona and a 7.3-m (24′) rigid-hulled Zodiac that we use mainly off Kaua'i and Ni'ihau. We have also moved these boats among islands for field projects between Kaua'i and Kona or worked with friends or collaborators who have boats on different islands.

Small boats have certain advantages over larger vessels for these surveys. We can go out on day trips and return to our own (rented) beds at night, we can eat out at restaurants, and when equipment breaks we can get replacements at a local store or on Amazon. The smaller boats are more maneuverable when working around animals, and we can choose when and where to go out, typically staying

Photographing a group of short-finned pilot whales near our research vessel off Kona, October 25, 2013. Much of our work has been concentrated off the Kona Coast, as the size of the island creates a large lee offering protection from the wind, allowing us to work in deep waters. Photo by author.

in much calmer waters where we can do a lot more with each group encountered. But small boats have disadvantages as well. They are generally restricted to calmer water, so our surveys have primarily been undertaken on the leeward sides of the islands—in Hawaiʻi, these are typically the west sides, with the winds coming from the east. The opportunities for us to work on the windward sides of the islands have been limited; it is mainly off Kauaʻi and Niʻihau, as well as Lānaʻi, where we've been able to work all around the islands. Thus much of what we know from our small-boat surveys is biased toward species and populations that might use the leeward sides of the islands differently than those that live on the windward sides, although we have some evidence that this is not the case for many species.

About half of all of our work has been off Kona, primarily because of the presence of very deep water close to shore there in comparison to the other islands, allowing us access to deepwater species. But it is also because there is a large lee off Kona—a calm area where it is easier to find animals and easier to work with them once we find them. Hawaiʻi Island is called the "Big Island" for a reason, and the size of the island itself, with its high mountains, creates a massive wind barrier, making the sea usually much calmer on the west side of the island. Next to Kona, our work has concentrated off Kauaʻi and Niʻihau, primarily because that is where the U.S. Navy does most of its training and testing activities, and we are trying to understand the populations that are exposed to naval sonar and what impacts it may have on different species of whales and dolphins.

We also take photos of whales and dolphins, often thousands of them in a day, primarily to identify individuals. Like humans, individual whales and dolphins are usually distinct, and as they get older they acquire more and more scars that make it easier to identify them from one encounter to another. We use different types of markings for identification purposes, depending on the species. Some, like rough-toothed dolphins or humpback whales, have distinctive pigmentation patterns that remain visible and stable for life. Others we identify using bite-wound scars, such as those from cookie-cutter sharks. These sharks are found in deep tropical waters and are perhaps best described as parasitic—they usually don't kill their prey, they just take somewhat circular or oval bites out of them, like a cookie punch. The bite wounds can be an inch or more deep, and when they heal the scars are often a different color than the rest of the body and may be visible for ten years or more. For most species, we use markings on the dorsal fin: through social interactions with others or attacks by predators, individuals often have nicks out of the trailing edge and sometimes the leading edge of the fin, the most easily photographed appendage of a whale or dolphin.

We have photo-identification catalogs of twelve species of whales and dolphins in Hawaiian waters, and we contribute photos to two other catalogs. For our catalogs, we assign each individual a catchy name, such as HIZc007, with "HI" representing Hawai'i, "Zc" representing the scientific name (in this case, *Ziphius cavirostris* for Cuvier's beaked whales). The "007" is a sequential number assigned to individuals as we add them to the catalog. Yes, it's not very imaginative, but

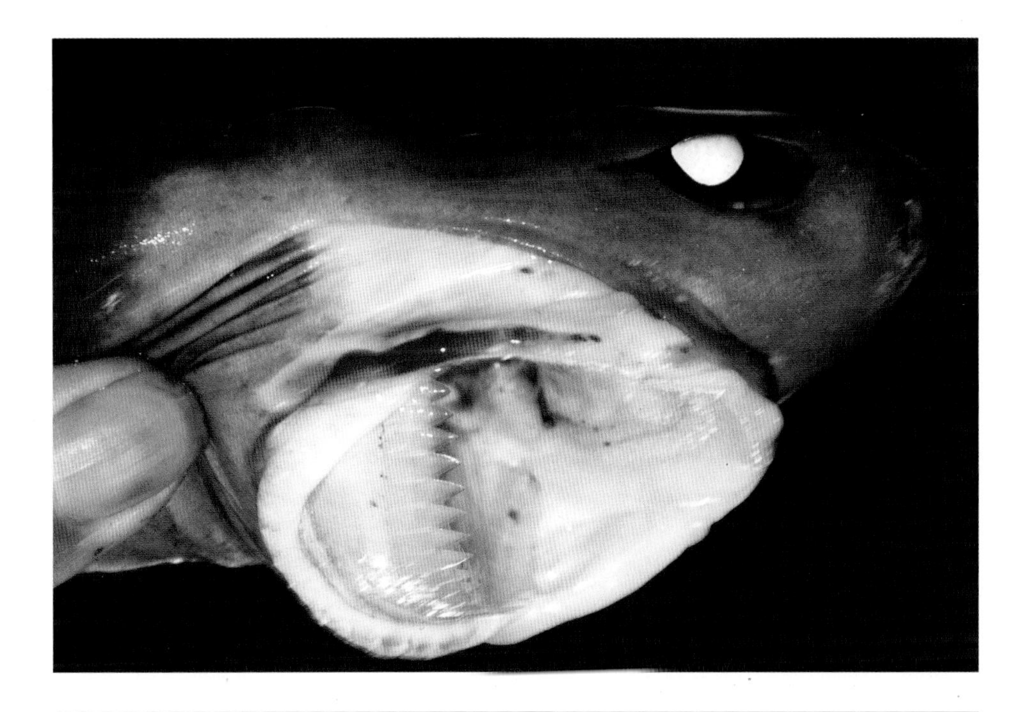

Closeup of the head of a cookie-cutter shark (*Isistius brasiliensis*) collected north of the main Hawaiian Islands in 2015. Photo by Jessica Chen.

An adult female Cuvier's beaked whale, HIZc007 in our photo-identification catalog, off Kona, November 21, 2014, with extensive white oval scars caused by cookie-cutter shark bites, as well as some linear scars caused by interacting with other whales. HIZc007 was first documented off Kona in 2004. Photo by Brenda K. Rone.

with over five thousand individuals of twelve species in different catalogs, it is hard to come up with good names for them all. That said, we have given a few individuals more familiar common names. One female false killer whale is called Mazie, named after the daughter of one of our catalog curators; one male is named Owen, after Sir Richard Owen, who first described false killer whales; and one is named Niho, Hawaiian for "tooth," as the individual has a broken tooth visible even with his mouth closed.

We use all these distinctive markings to track individuals over time and space—we can find out whether individuals have constant companions or best buddies, or if they just mix and match with others randomly. Based on how often we resight individuals compared to identifying new ones, we can also estimate abundance using what are called mark-recapture methods. For small populations with high resighting rates of individuals, these methods are typically much better than the line transect methods from large-vessel surveys, and they produce estimates that have a lot less uncertainty. Movements can also be documented, either among islands as, for example, with false killer whales, or between Hawai'i and various feeding grounds elsewhere, as with species such as humpback whales or North Pacific right whales.

We are also making efforts to measure individuals in the wild to help assess the relative age of animals (adults versus juveniles). One of the cameras we use has a system with two lasers mounted on the lens: these lasers project green dots onto an animal precisely 15 centimeters (cm) apart. If the individual is parallel to the camera, we can use this to measure the height and length of the dorsal fin and often the length from the dorsal fin to the blowhole. We can compare these measurements to known aged animals (typically from stranded individuals) to determine whether a female might be sexually mature, for example.

Like the small boat launches off large-vessel surveys, we are also collecting biopsy samples, and we've made efforts over the years to collect samples from individuals of as many species as possible off each of the island areas. These samples, combined with those collected offshore and in other areas, have been the basis for genetic analyses of many different species, both to determine the sex of the individuals and also to assess whether there is only a single population of each species in Hawaiian waters or many genetically isolated or distinct populations. We've provided these samples to collaborators with the NMFS

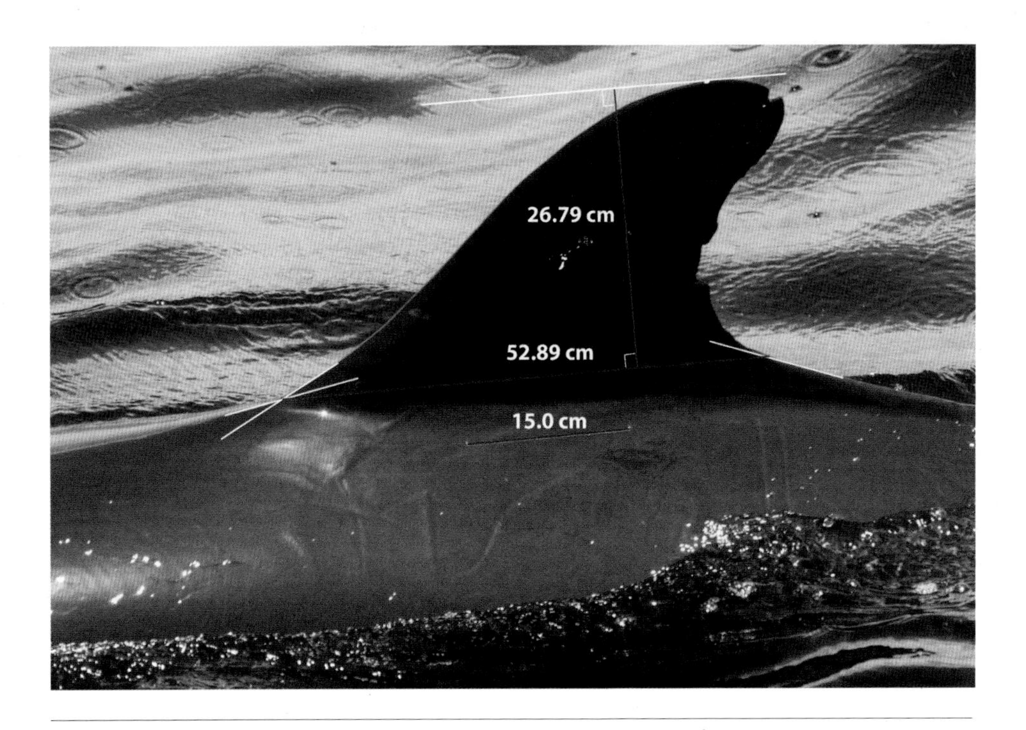

Using a pair of green dot lasers mounted exactly 15 cm apart, we are able to measure the dorsal fin and other parts of the body of different species, to help determine growth rates and estimate relative age of individuals. This is a false killer whale from the main Hawaiian Islands population, August 20, 2011. Photo by Elisa A. Weiss.

Southwest Fisheries Science Center, as well as Portland State University, Oregon State University, and Scripps Institution of Oceanography, for genetic analyses of a variety of species. It turns out that for every species so far examined in detail—those for which we've been able to collect enough samples—there is evidence of multiple populations in Hawaiian waters.

Since 2006, we've also been using remotely deployed satellite tags to study the movements and habitat use of a number of species. These tags were first developed by Russ Andrews of the Alaska SeaLife Center, one of our collaborators. Greg Schorr from Cascadia has helped test and refine these tags, their anchor

LIMPET satellite tags are remotely deployed using an air rifle and allow us to track movements over periods of weeks to months. This photo shows an adult female sperm whale immediately after being tagged off Kona, October 31, 2013. The tag is visible on the middle of the fin, while the tagging arrow has bounced off and is above the whale's fin. Photo by author.

mechanisms (two surgical-grade titanium darts), and the deployment system—we use an air rifle—so that we can reliably deploy the tags on a wide range of species. The tags are expensive, from $3,500 to $5,000, and can only be used once, so having good marksmen is key: both Greg and Daniel Webster from Cascadia have now deployed more than 270 tags on twelve species of whales and dolphins in Hawaiian waters, while only losing a few. The tags adhere to the dorsal fin and remain attached for weeks or months. Information on movements and (for some tags) diving behavior is obtained from receivers on orbiting weather satellites. These tags have helped deal with the bias of working primarily on the leeward sides of the islands and typically relatively close to shore, as once they are tagged, individuals are free to go wherever they want, around the islands or offshore. As with biopsy sampling, when animals are struck with these dart tags, they typically flinch and dive, but they usually return to their normal behavior within minutes, and most are as approachable afterward as they were before.

We also record a great deal of other information when we are with a group: the group size, the number of adult males (when that is obvious), how many newborn individuals there are, whether other species are present and how are they interacting, and many aspects of their behavior. When on the water, we also record other creatures—how many flying fish we see, the number and types of seabirds, the number of game fish—and even the number of fishing vessels seen. All of these things tell us something about what is going on in the environment in which the different species live. Many of the species we study eat squid, and we also collect squid that are found floating dead at the surface to get a better idea of what types of prey are in the area, and we make acoustic recordings of some species to help fill in gaps for acousticians who work with recordings from autonomous systems.

Information that comes from examination of stranded animals also fills a lot of gaps. Most of these are individuals that were sick and have come into shore to die, and they are either found dead or die shortly after being found. Researchers

›››

An adult male killer whale that live stranded on Kaua'i, October 22, 2008. The necropsy results showed the individual was severely emaciated. Photo by Jessica M. Aschettino.

Hawai'i Pacific
University
STRANDING
TEAM

with Hawai'i Pacific University respond to strandings of whales and dolphins, conducting necropsies on dead animals both to try to find out why they died and to collect information that may be useful for understanding the species or the populations here in Hawaiian waters. This includes examining teeth for aging, samples for genetics and toxicology, stomach contents to determine diet, and reproductive tracts (ovaries or testes) to determine whether an individual is sexually mature. All strandings, dead or alive, dead animals offshore, or animals in distress should immediately be reported to the National Marine Fisheries Service's toll-free twenty-four-hour hotline (1-888-256-9840); they coordinate responses to strandings. If small animals, such as dolphins, are found dead floating offshore, I recommend bringing them into shore if no one can be reached by phone to ask. In cases where they've been left offshore, they are typically consumed by sharks quickly and are never found again.

It is a combination of approaches—behavioral observations, photo-identification, genetic analyses, acoustic recordings, examination of stranded animals, and satellite tagging—that really gives us the most complete picture of the lives of whales and dolphins in Hawaiian waters. For every day or week we spend on the water, there may be enough information obtained for months of analyses by folks in the Cascadia office or by our collaborators scattered throughout many different research institutions.

I live in Olympia, Washington, but make trips out to Hawai'i about four times a year for field projects, typically working from two or three different islands in any one year. The trips may last from eight days to three or four weeks. During these trips, we assemble a team and work with local researchers and others who spend time on the water in Hawai'i: whale-watching operators and naturalists, photographers, educators, fishermen, and students from local universities. These trips are an opportunity for us to collect information for later scientific publications, conferences, and workshops and also an opportunity to learn from and teach others. Slowly we are whittling away at some questions and asking new ones, as we gain a more sophisticated understanding of the different populations and social groups of whales and dolphins in Hawaiian waters. Although we've made tremendous progress, there is clearly still much to be learned about most species of Hawaiian whales and dolphins, as well as how human activities may be influencing them.

SPECIES ACCOUNTS

Accounts for each species follow, with an introduction to each family or superfamily of whales or dolphins found in Hawaiian waters, as well as introductions to some of the subgroups (for example, the blackfish). I make no attempt to present the same level of detail for each species but instead provide more information for those that are seen relatively frequently in Hawai'i or those whose populations are better known in Hawaiian waters than almost anywhere else in the world. The order of species presented also differs from traditional field guides for whales and dolphins. I start with the oceanic dolphins, as they include the species seen most often, and I leave the baleen whales for last, as only one of the seven species is regularly found near shore around the main Hawaiian Islands. Within the oceanic dolphins, the presentation of species starts with the blackfish, both because they are my favorite group and because the best-known populations of several of these species anywhere in the world are found in Hawai'i. The maps included show actual sighting records and, for those species that we have satellite tag data, include the tracks or locations of tagged individuals to help illustrate movements or lack thereof and to better represent where the animals spend their time. Sighting records come from published sources for some rarely seen species (e.g., killer, sei, minke, Bryde's, and North Pacific right whales), about 800 records from NMFS surveys in Hawaiian waters since 2002 and another 800 records from various friends, colleagues, and members of the public, and over 2,500 records from our own work. Most of the abundance estimates presented come from the 2002 or 2010 NMFS surveys, but some come from mark-recapture analyses of photographs. Photos for each species are included; most are from Hawaiian waters, but when no suitable photos were available from Hawai'i, I have substituted photos from elsewhere and noted the location.

THE OCEANIC DOLPHINS

The family Delphinidae, the oceanic dolphins, is the largest family of cetaceans, with thirty-eight species and counting. By the time you read this, there may be thirty-nine or forty species recognized, as advances in genetics and a rapid increase in marine mammal research around the world are revealing previously unrecognized species. This group includes both truly oceanic species and some that live in very shallow nearshore waters or even in rivers. It is an incredibly diverse group, including some of the best-known species of toothed whales in the world, such as killer whales and the ubiquitous common bottlenose dolphins, as well as

Members of the family Delphinidae, the oceanic dolphins, include five species referred to as "whales," such as these short-finned pilot whales. This photo was taken off Kona, March 22, 2015. Photo by Deron S. Verbeck/iamaquatic.com.

some poorly known species, such as southern right whale dolphins and Heaviside's dolphins.

Almost a third of oceanic dolphin species worldwide have been documented in Hawaiian waters: twelve species representing ten different genera. There is also an additional species that normally lives a bit farther north, which is likely to be recorded at some point in Hawaiian waters. Five of the delphinids found in Hawaiian waters are commonly referred to as "whales" rather than "dolphins," but they are more closely related to the rest of the delphinids than they are to other whales. The terms "whale" and "dolphin" can definitely be misleading. These names were applied a long time ago, before there was a good idea of the relationships among different species, and it wasn't size that determined whether a species was called a whale or a dolphin. The so-called whales include three of the largest delphinids (killer whales, short-finned pilot whales, and false killer whales), as well as two of the smaller species (melon-headed whales and pygmy killer whales). The Risso's dolphin is one species of delphinid in Hawaiian waters that is larger than either the melon-headed whale or the pygmy killer whale. The names become even more confusing, as the term "porpoise" is often used by fishermen in Hawaiʻi to refer to any of several small species of dolphins, although technically it refers to a group of small cetaceans in the family Phocoenidae, none of which are found in Hawaiian waters.

Of the twelve species of oceanic dolphins in Hawaiian waters, there are eight with known resident, island-associated populations: false killer whales, short-finned pilot whales, pygmy killer whales, melon-headed whales, rough-toothed dolphins, common bottlenose dolphins, spinner dolphins, and pantropical spotted dolphins. Five of these have had multiple stocks recognized in Hawaiian waters by the National Marine Fisheries Service, either based on genetic analyses or on a combination of genetics and results from photo-identification and satellite tagging studies. There are four species of delphinids that have no known resident populations around the islands: killer whales, striped dolphins, Risso's dolphins, and Fraser's dolphins. None of these are likely to have resident populations, given their low sighting frequencies and offshore habits.

In Hawaiian waters, these twelve species range from the extremely rare (pygmy killer whales, killer whales, and false killer whales) to the superabundant (Fraser's dolphins, pantropical spotted dolphins, striped dolphins, and rough-toothed dolphins). Only one stock among the twelve species has been listed under the U.S. Endangered Species Act: the insular population of false killer whales in the main Hawaiian Islands.

THE BLACKFISH

In his iconic novel *Moby Dick,* Herman Melville refers to the "Black Fish" as the "popular fishermen's names for all these fish, for generally they are the best." The term "blackfish" has been used by fishermen at least since the seventeenth century to describe pilot whales and several other similar species. Five species of oceanic dolphins in Hawai'i have often been referred to in this way, and it applies for obvious reasons—all five species are largely black in color, at least on the sides and dorsal surface. These five are the delphinids also referred to as whales—false killer whales, short-finned pilot whales, killer whales, melon-headed whales, and pygmy killer whales—and they are related to each other. They all have a rounded head, lacking a distinct rostrum (beak or snout), a trait shared with another closely related species, the Risso's dolphin.

While killer whales are easy to recognize, with their striking black and white coloration, the other four are easily and often confused, both by the general public and at least sometimes by scientists. Two of the four species are relatively large (false killer whales and short-finned pilot whales), while two are relatively small (pygmy killer whales and melon-headed whales), so size alone cannot be used to tell them apart. Without photos, the best way to tell these species apart comes from paying attention to subtle details of body proportions (for example, the dorsal fin size relative to the amount of back visible) and coloration—yes, they are all largely black, but some do have lighter gray sides. The shape of the darker dorsal cape and the distinctiveness of the border of the cape are also important features to note.

While killer whales are distributed worldwide and are well known to both the general public and the scientific community, the other four species, which I refer to as Hawaiian blackfish, tend to live in deeper water farther from shore, primarily in tropical and warm temperate areas, where they are not as easily accessible either to the general public or to scientists. With the deep and somewhat protected waters close to shore facilitating research, more is known about three of these species in Hawai'i than anywhere else in the world, and the Hawaiian population of the fourth is a close runner-up.

FALSE KILLER WHALES *(Pseudorca crassidens)*

In the early 1960s, commercial longline fishermen off Kona reported "pilot whales" taking fish off their lines. Intrigued by the fishermen's reports, the collector for Sea Life Park, which opened on O'ahu in October 1963, went out on one of the longline vessels that year and found that the small black whales were actually false killer whales.

In 1984, researcher Dan McSweeney was following a group of false killer whales off the Kona Coast and slipped into the water with a mask, a small scuba tank, and an underwater camera. Two black shapes moved by below, vocalizing. Dan turned, and a third individual was swimming rapidly toward him, carrying most of a large *'ahi,* a yellowfin tuna, weighing over 45 kilograms (kg) (100 pounds [lbs]). The whale stopped a couple of meters away and opened its mouth, letting the fish go, and the momentum carried the fish toward Dan. The whale was obviously offering the fish to him, and Dan reached out and took it. The false killer whale started blowing bubbles, moved away, then turned rapidly and came back, stopping next to him again. Dan pushed the fish back toward the whale; it took it slowly and deliberately, then moved away and joined its companions. The whales passed the fish back and forth and started to consume it, and all had a share.

Such prey sharing, among themselves and even occasionally with humans, is one of the most unusual aspects of the behavior of false killer whales. A lone individual in British Columbia, Canada, in the late 1980s, far from its normal range, would approach boaters and offer them salmon. In Hawai'i, I've witnessed them catching large fish such as *ono* or *mahi mahi,* then passing them back and forth among themselves without any of them taking a bite, before the fish is returned to the individual that caught it, and then all begin to share. Such behavior probably serves to reinforce the strong bonds among individuals that may be constant and long-term hunting companions in an environment where the benefits of cooperatively finding and catching prey allow them to survive as top predators.

False killer whales have always been our highest priority species when encountered. We see them rarely, but when we do we try to make the most of the opportunity, photographing all the individuals present, recording information on prey captured, collecting biopsy samples that are used to study genetic relationships

and toxin levels, and deploying satellite tags to examine movements. Most false killer whales have distinctive markings on the dorsal fin, allowing us to follow individuals over time and space. So although we don't see them often, they are one of the best-known species of whales or dolphins in Hawaiian waters. In particular, the population in the main Hawaiian Islands is the best-known of this species anywhere in the world. This population was listed as endangered under the Endangered Species Act in 2012, and it is the single whale or dolphin population most at risk in Hawaiian waters.

Identifying Features and Similar Species

Born at about 1.5 to 2.1 m (4′11″ to 6′11″) in length, the longest false killer whale recorded was 5.96 m (~19′7″). They are known to vary in length between populations, and how the Hawaiian animals compare is unknown, as only a few have ever been measured. Two known individuals from the main Hawaiian Islands population that stranded in recent years, one a twenty-four-year-old female and the other a twenty-two-year-old male, measured 4.2 m (13′9″) and 4.45 m (14′7″), respectively. False killer whales don't stop growing until they are between twenty-five and thirty years of age, so neither of these individuals was probably full size.

They are a dark gray in color, often appearing black on all but a small part of the ventral (lower) surface, where they have a pale blaze between the pectoral fins, a pale stripe running down the middle of the belly, and a wider pale area in the genital area. In good lighting conditions, the slightly darker gray dorsal cape can be seen, with a diffuse boundary between the cape and the lighter gray sides. They are bitten by cookie-cutter sharks, but the scars heal to the same color as the background. On average, adult males are 0.7 m longer than adult females, and they also differ in head shape—the front of the head of an adult male overhangs the lower jaw to a greater extent and in older males is flattened. Their dorsal fin is

‹‹‹

False killer whales sharing an *'ahi*, a yellowfin tuna, off Kona, 1984. Photo © Dan J. McSweeney/Wild Whale Research Foundation.

A subadult false killer whale from the main Hawaiian Islands population off Kona, March 12, 2012. Photo by Deron S. Verbeck/iamaquatic.com.

A mother and calf pair of false killer whales from the main Hawaiian Islands population, off O'ahu, October 22, 2010. The mother, HIPc116 in our catalog, was first documented off Kona in 1990. Photo by author.

located in the middle of the back and generally curves backward, but in Hawaiian waters there is a lot of variability in dorsal fin shape, often caused by injury. In Hawai'i there is one individual with a triangular fin, one missing the dorsal fin entirely, and three individuals with dorsal fins collapsed over to the side. On a normal surfacing, their dorsal fin is smaller in proportion to the amount of back visible than any of the other blackfish. They are the most aerial of the Hawaiian blackfish, often coming completely clear of the water, particularly when attacking prey. When attacking large fish, false killer whales will often leave the water with the belly upward, as they are trying to strike the fish with the underside of

Adult male false killer whales (top) are larger than females (middle), and their rostrum overhangs the mouth to a greater extent. Illustrations by Uko Gorter.

their tail flukes. None of the other Hawaiian blackfish do this, so such behavior, in combination with other identification features, can be used to confirm species. They are the only species of Hawaiian blackfish to throw fish, particularly *mahi mahi,* quite high in the air like this.

As mentioned, there are three similar-appearing species, and all overlap to some degree in size. Short-finned pilot whales overlap almost completely in size, while large melon-headed whales and pygmy killer whales are about the same size as juvenile false killer whales. If all that is visible is the dorsal fin and back, the most obvious way of telling false killer whales from the other species is the

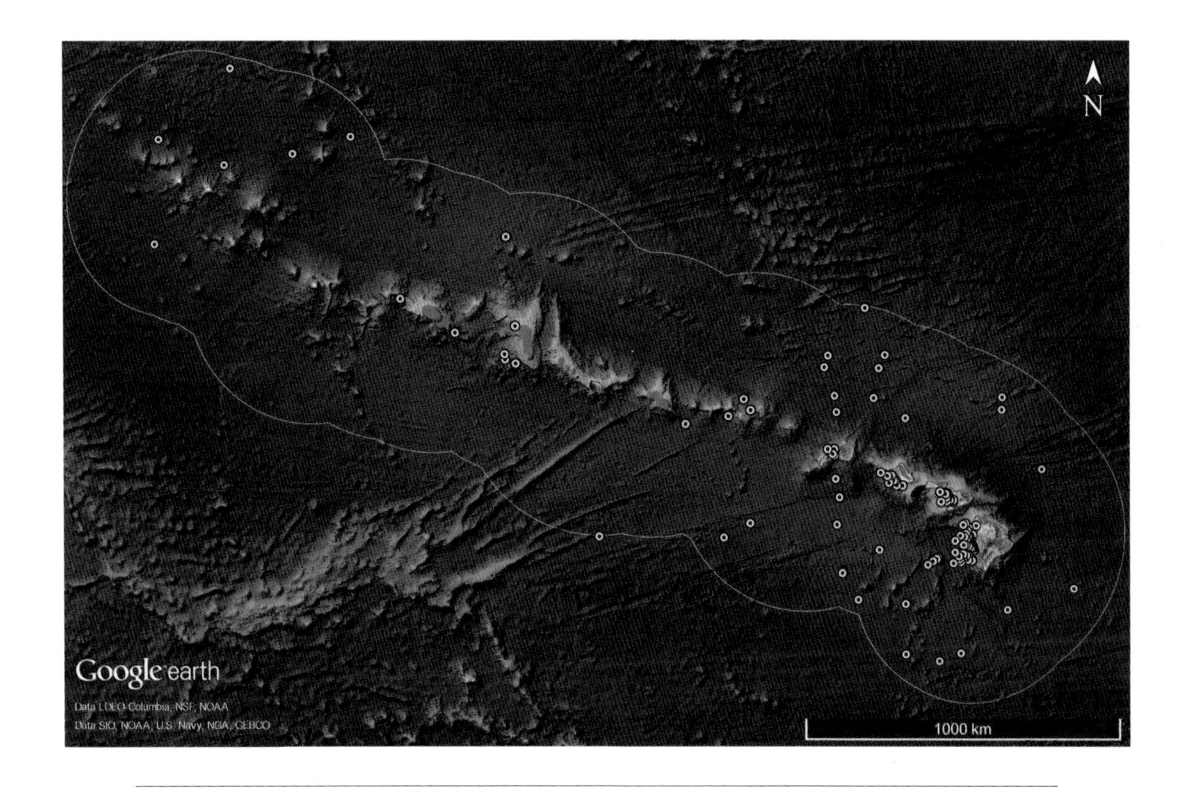

Sightings of false killer whales in Hawaiian waters show they are distributed offshore and around the islands, although the distribution varies among the three populations documented in Hawaiian waters.

size of the dorsal fin relative to the amount of back showing: false killer whales have the smallest fin in proportion to the amount of body visible. Short-finned pilot whales have a dorsal fin that is longer than it is tall, and the fin is positioned in front of the midpoint of the body. False killer whales have an unusually shaped pectoral flipper that can be used to tell them apart from other species: the leading edge has a distinct bulge, giving it an S-shaped appearance. There are behavioral differences as well: melon-headed whales, pygmy killer whales, and short-finned pilot whales will all rest during the day by floating motionless at the surface, whereas such behavior is extremely uncommon for false killer whales.

Habitat Use, Movements, and Abundance

False killer whales are distributed in tropical and warm temperate waters worldwide, but they are primarily found in deeper open-ocean areas. They are top predators, and top predators are naturally rare—nowhere throughout their range are they particularly common. False killer whales are one of only two species subjected to intensive studies in the Northwestern Hawaiian Islands (the other being spinner dolphins), with two individuals satellite tagged there in 2010 and another two satellite tagged there in 2013, both part of NMFS cruises. That work helped confirm that there are three different partially overlapping populations in Hawaiian waters, and each is genetically distinct from the others. The smallest of these is the main Hawaiian Islands insular population, estimated at between 150 and 200 individuals, which ranges from Ni'ihau to the east side of Hawai'i Island and as far offshore as about 120 km. There is another island-associated population that lives in the eastern half of the Northwestern Hawaiian Islands and overlaps with the main Hawaiian Islands population around Kaua'i and Ni'ihau. This population is thought to be about three times the size, numbering about 600 individuals. There have been two sightings of individuals from this population as far east as Wai'anae, O'ahu, one in April 2013 and one in June 2015. Tagged animals from this population have moved only as far west as Gardner Pinnacles, to the west of French Frigate Shoals; whether they range as far as Midway or Kure Atolls will require additional satellite tagging to confirm. In offshore waters, there is an open-ocean population that roams widely and is much larger than the two island-associated populations combined. Not surprisingly, the two island-associated populations tend to use shallower water than the offshore population—we've seen individuals from the former in water as shallow as about 50 m. But from satellite tagging data, they also use depths out to almost 5,000 m, and the average depths used are in the range of 500 to 1,500 m. Individuals from the offshore population are found in waters with average depths in the 4,000 to 5,000 m range, although they too can be in shallower waters on occasion, as shallow as 1,000 m deep.

Movements of individuals and the overlapping ranges of the populations have been based in part on photo-identification data. Almost all of the individuals in the main Hawaiian Islands population are distinctive and have been identified.

Locations of satellite tagged false killer whales from the endangered main Hawaiian Islands population, obtained from thirty-three individuals tagged between 2007 and 2014. The dense clusters of points off the north end of Hawai'i Island, north of Maui and Moloka'i, and to the southwest of Lāna'i all represent high use areas for this population.

Thus if photos are obtained of three or more individuals from a group of false killer whales in Hawaiian waters, if they are part of the main Hawaiian Islands population, chances are we'll be able to recognize at least one or two of them. Smaller numbers of individuals from the other two populations have been photo identified, and there have been a number of resightings of individuals from the Northwestern Hawaiian Islands population off Kaua'i and O'ahu. We know both from photos and from satellite tag data that individuals move rapidly among the

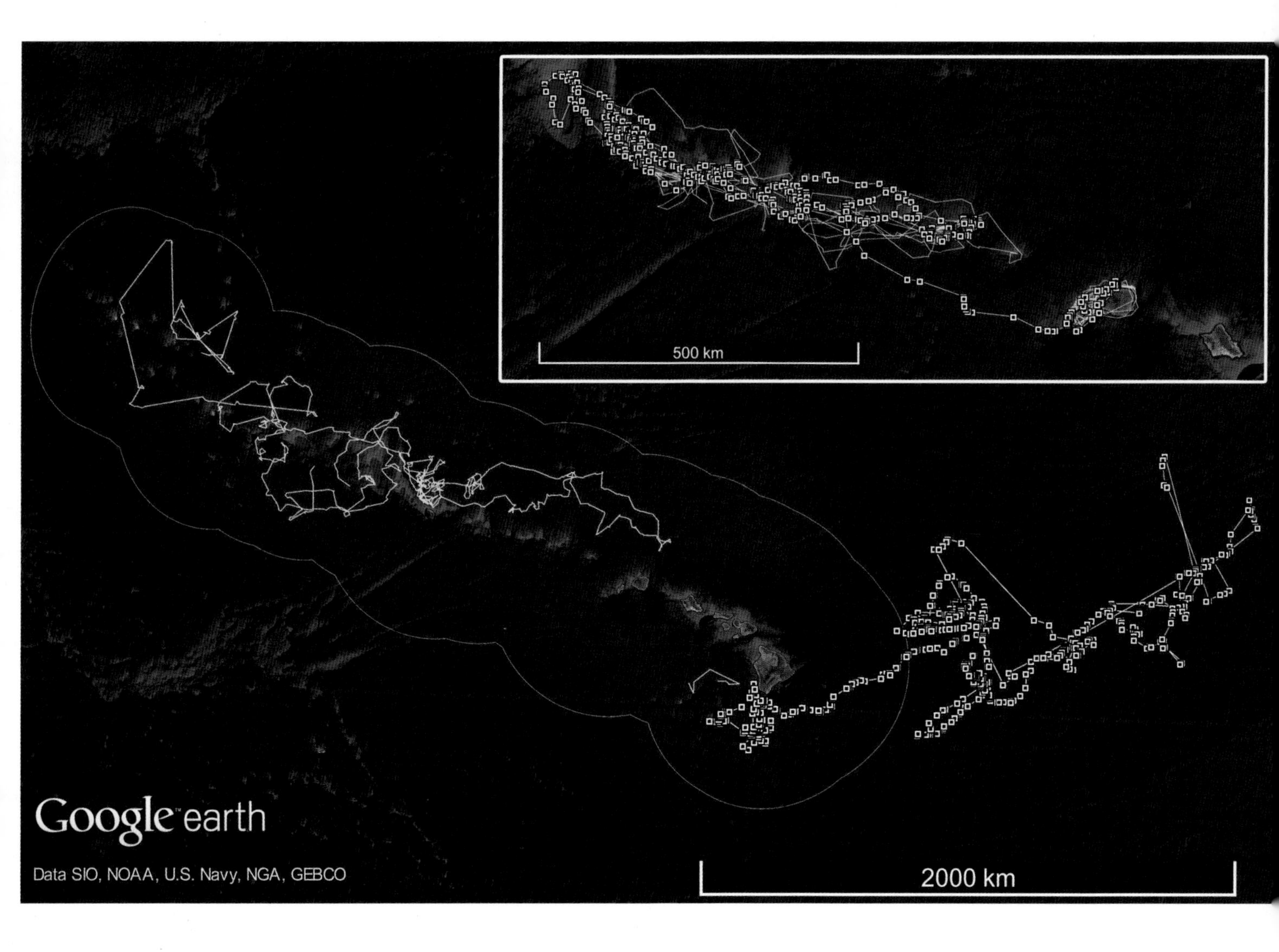

Tracks of six satellite tagged false killer whales from the pelagic population, tagged in 2008 and 2013, showing broad-ranging movements inside and outside the U.S. Exclusive Economic Zone surrounding the Hawaiian Islands. Locations for one individual over a 122-day span are shown with white squares, and consecutive locations are joined with a line. The inset shows tracks (lines) from six satellite tagged false killer whales from the Northwestern Hawaiian Islands population tagged in 2010, 2012, and 2013 and locations (white squares) over a 42-day period for one whale tagged off Kaua'i in June 2012.

islands, and their overall movement rates are faster than any of the other species we've satellite tagged. They regularly move back and forth from the windward (northern and eastern) to the leeward (southern and western) sides of the islands, and they spend equal amounts of time on the windward and leeward sides of the islands. But when on the windward sides, they concentrate closer to shore, heavily using areas in the 500–1,200 m depth range, particularly north of Hawai'i Island, Maui, and Moloka'i. While on the leeward sides of the islands, individuals from this population tend to spread out over much larger areas, both near and far from shore.

Predators and Prey

Although it is unlikely to happen in Hawai'i given the rarity of both species, one killer whale attack has been documented on a group of false killer whales off New Zealand. Large sharks such as tigers or whites do occasionally attack false killer whales; we have one adult female and one juvenile in our photo-identification catalog from the main Hawaiian Islands population with healed bite wounds from large sharks.

False killer whales hunt and travel in a loose pack, like wolves, actively hunting during the day. Their prey are often near the surface, and when a kill is made other members of the group converge and the catch is shared among them. For all of these reasons, it is easier to observe what they are feeding on than for any other species of whale or dolphin in Hawaiian waters. They primarily feed on large game fish, including *mahi mahi, ono, 'ahi, 'ahi pālaha* (albacore tuna), *aku* (skipjack tuna), and *a'u ku* (broadbill swordfish), and they have also been seen feeding on *loulu* (scrawled filefish), *monchong* (lustrous pomfret), and *ulua kihikihi* (threadfin jack). One stranded individual, an adult female from the main Hawaiian Islands population, also had a number of squid in its stomach. Although it has been reported that false killer whales have killed a humpback whale calf in Hawaiian waters, I spoke with one of the individuals present for the "attack," and it sounds like the evidence was circumstantial at best.

False killer whales have been seen to attack sperm whales off the Galapagos, although the attack wasn't fatal. The researchers who witnessed the attack thought that it may have been a form of kleptoparasitism (prey stealing),

A false killer whale eating a *mahi mahi,* the most frequently documented prey species for false killer whales from the main Hawaiian Islands population. Photo © Dan J. McSweeney/Wild Whale Research Foundation, off Kona, October 22, 2005.

A false killer whale eating an *ono*, or wahoo, off Kona, December 11, 2010. Photo by Daniel L. Webster.

forcing the sperm whales to regurgitate so the false killer whales could consume the regurgitating squid. False killer whales have been observed attacking dolphins that had been captured in the tuna purse seine fishery in the eastern tropical Pacific (ETP) as they were released from the net, typically stunned and disoriented. I think this reflects more the ability of false killer whales to learn to feed on new things when the opportunities arise, like taking fish from longlines, than it does any particular tendency of false killer whales to feed on other dolphins. In Hawai'i we often see other species of dolphins with false killer whales, including bottlenose dolphins, rough-toothed dolphins, and even spinner dolphins on one occasion, and we've never seen any potential aggression toward them. We have seen them swim by humpback whales while apparently ignoring them, although the humpbacks appeared agitated when the false killer whales came close.

Life History and Behavior

False killer whales, like their close relatives the killer whales and pilot whales, are long lived and slow to reproduce. The oldest animal aged, using the traditional method of sectioning the teeth, was a female sixty-three years of age, although it is likely that some individuals may live longer, and it wouldn't surprise me if individuals in their seventies or eighties are eventually documented. While females mature earlier than males, giving birth to their first calf between nine and twelve years of age, they also live longer than males. Males don't become sexually mature until their middle to late teens, and the oldest known male was fifty-eight years of age.

One of the most unusual aspects of their life history is that females go through menopause, like killer whales, pilot whales, and humans, and then live for another ten or twenty years. Female false killer whales stop reproducing by about forty-four years of age. Menopause is unusual in the animal kingdom and is thought to have evolved because older females perform a more important role as a grandmother or auntie than they would by having more calves themselves. In the case of false killer whales, like killer whales, it is likely that older females carry most of the cultural knowledge, such as knowing the best foraging areas when environmental conditions are unusual, such as during El Niño events. Females might also lead the group or coordinate the group when cooperatively hunting. Females give birth to a single calf only every seven or so years, so they are likely to have only four or five calves over their reproductive life span. There is no information on mortality rates of newborns, but for the closely related killer whale, mortality in the first year of life may be close to 50 percent, and it also likely high for false killer whales. Thus, of the resident species of whales and dolphins in Hawaiian waters, false killer whales, along with short-finned pilot whales, likely have the lowest reproductive potential.

One of the most distinctive false killer whales from the main Hawaiian Islands population, HIPc197, first documented off Kona in 1986. HIPc197 is a member of Cluster 2, a rarely seen group from this population. This photo was taken off Kona on August 13, 2010. Photo by Jessica M. Aschettino.

Species of toothed whales that share these types of life history traits—long lives, long calving intervals, delayed maturity—all tend to be similar in that they have very strong and enduring bonds among individuals. From analyses of association patterns of individuals in the main Hawaiian Islands false killer whale population, there are three main social groups. We call them "clusters," but they are effectively the same as "pods" of killer whales or pilot whales. Scientists tend not to be very creative in how they name such groups, so we call them Cluster 1, Cluster 2, and Cluster 3. Actually, we've also determined that there are two sub-clusters in Cluster 1, which we've affectionately named Sub-cluster 1A and Sub-cluster 1B. These clusters appear to be long-term stable groups, made up primarily of closely related individuals. Some of the bonds between individuals likely last for life, and females appear to remain within the social group in which they were born. Paternity analyses indicate that mating occurs both within and between these social clusters. Some males have all the luck—one male in particular was responsible for seven different offspring.

Group sizes that we encounter in Hawai'i are quite variable, from a pair of individuals up to more than forty. One of the common features of our encounters is that when we find one group of false killer whales, it is usually part of a much larger aggregation of individuals, with scattered subgroups over a very wide area. One aggregation had subgroups spread over more than 20 km. We've seen such subgroups splitting and merging from satellite tag data as well. In 2008 we tagged five different individuals in the same group, and they repeatedly joined and spread apart over 50 km or more as they roamed around the island of Hawai'i, searching for prey.[1]

Conservation

Over the years, false killer whales in Hawai'i have been affected by humans in a variety of ways. In the 1960s and 1970s, about a dozen false killer whales were captured in Hawai'i for captive display. In the late 1980s the number of U.S. longline vessels fishing in Hawai'i increased from less than 40 to about 140, and it wasn't until 1992 that longline fishing near shore was prohibited. Today they face a variety of threats, and in 2012 the main Hawaiian Islands insular population of false killer whales was listed as endangered under the U.S. Endangered Species Act. Other than sperm whales, which were listed as endangered in 1970 due to commercial whaling, this is the only species of odontocete in Hawaiian waters listed as endangered, and it is certainly the most threatened population in Hawaiian waters. They were listed for very different reasons than were sperm whales or the larger baleen whales, all of which were listed because of depletion from commercial whaling, which ceased many years ago.

1. An animation of the movements of these five individuals over a ten-day period is available at http://www.cascadiaresearch.org/Hawaii/falsekillerwhale.htm.

False killer whales in Hawai'i face a number of threats, many of them insidious—they are difficult to observe and difficult to do much about. The most subtle threat comes from the impacts of persistent organic pollutants such as pesticides and industrial chemicals; for example, dichlorodiphenyltrichloroethane (DDT), polychlorinated biphenyls (PCBs), and flame retardants. These chemicals are all lipophilic—that is, they are attracted to fats, and thus they get stored in the blubber of whales and dolphins. Because false killer whales are at the top of the oceanic food web, feeding on various predatory fish that are accumulating toxins, they have levels higher than any other species except perhaps killer whales. Once these toxins reach certain levels, they can suppress the immune system or compromise reproduction. The high levels don't kill animals outright, but they can increase the likelihood of death from what otherwise might be nonfatal diseases or infections. Because these toxins are lipophilic, when females reproduce and nurse their offspring with fat-rich milk, their levels of these toxins decrease dramatically, while the levels in their dependent calves can become very high very quickly. This is particularly the case for the firstborn calf of any female, as she may have accumulated ten or more years of toxins that are transferred to her first calf, while the shorter time intervals between subsequent calves mean there is less time to accumulate high levels of toxins. Analysis of toxin levels in biopsy samples we've collected show that every adult male false killer whale and many of the adult females have levels of PCBs that are high enough to compromise their health.

Most of the fish species that false killer whales feed on are ones that humans like to eat as well, and this leads to two different conservation issues. Some of these fish, in particular *'ahi,* are a lot smaller today than those that were caught sixty or eighty years ago, and there are a lot fewer of them. The average weight of an *'ahi* caught today in Hawaiian waters is about half that of those caught in the 1940s. The big fish are the target of both recreational and commercial fisheries, and, at least in the case of *'ahi,* these fisheries have had a long-term impact, both on the size of the fish and their numbers. What impact this may have had or is having on false killer whale populations is hard to say, but with fewer fish and smaller fish, individuals may have to spend more time or more energy hunting and devote less energy to reproduction. Interestingly, *mahi mahi* have increased in abundance in the central tropical Pacific in the last thirty years, probably as

When being chased by false killer whales, *mahi mahi* will often hide under boats or next to divers or snorkelers. This individual was hiding behind the photographer, but this did not deter the false killer whale from coming in to take the fish! The *mahi mahi* had already been injured by the whale and tooth rakes are visible along the side. Photo by Deron S. Verbeck/iamaquatic.com off Kona, March 17, 2012.

a result of reduced competition by *'ahi* and other large game fish. False killer whales do eat a lot of *mahi mahi,* so it is possible that their increase may have compensated for the reduction in the size and number of *'ahi.*

Sharing their prey is an integral part of the life of a false killer whale, so when humans offer fish to them, at the end of a fishing line, I don't think there is any malicious intent on the part of the whale to "steal" the fish, although that is not how it is perceived by most fishermen. False killer whales have been taking fish off fishermen's lines in Hawaiian waters for a long time, and such depredation, as it is called, can cause problems for the whales in several ways. A male false killer whale from the main Hawaiian Island population was found dead in October 2013 at Ka Lae, Hawai'i Island. When the animal was examined, he had five hooks in his stomach; based on the types of hooks, these came from a number of different fisheries. This individual did not die from ingesting the hooks, but more often than not when a whale or dolphin swallows a hook, it does lead to death. How often individuals from the main Hawaiian Islands population take fish or get hooked is unknown, but ingesting fishing gear certainly has the potential to kill individuals. In the offshore longline fishery for *'ahi po'onui* (bigeye tuna), false killer whales are the most frequently hooked cetacean species, and the bycatch in that fishery, primarily individuals from the offshore false killer whale population, is higher than the population can sustain. I suspect that false killer whales are also at least occasionally shot at by some fishermen out of retaliation or to try to deter them from taking their catch.

SHORT-FINNED PILOT WHALES *(Globicephala macrorhynchus)*

Short-finned pilot whales were first documented in Hawaiian waters in the late 1950s, with four different mass strandings recorded on Oʻahu, Lānaʻi, and Kauaʻi between 1957 and 1959. Altogether, they totaled sixty-six individuals, most of which died.

When working off Maui in 2000 and 2001, the deep waters to the west of Lānaʻi seemed a long way away given the size of the boat we were using and our limited fuel budget. But when we did make it out there, we saw short-finned pilot whales almost every day, a species that did not venture into the waters between Maui and Lānaʻi. We collected biopsy samples for later genetic studies and photos to establish a catalog of recognizable individuals. It wasn't until our study area expanded to include Hawaiʻi Island in 2002 that sightings of short-finned pilot whales became commonplace. On our first day on the water off Hawaiʻi Island in April 2002, our third encounter of the day was a group of twenty short-finned pilot whales, and we deployed a suction-cup-attached time-depth recorder on an adult male to study diving behavior. We had to leave the group early due to engine problems, but the next day, with the help of a VHF radio transmitter on the tag, we were able to find the tag floating offshore and successfully download the data. The tag had remained attached for twenty-two hours, giving us a detailed look into the day- and nighttime patterns of a short-finned pilot whale for the first time. In the two hours after we tagged the whale, it remained close to the surface, never diving below 20 m and moving slowly; the tag also had a paddlewheel sensor that measured swim speed. Then at 3:30 in the afternoon, the whale began a series of deep dives, some to over 700 m, repeating until just after sunset, when he started diving to just 200 to 400 m. Around 4:00 a.m. the next day, the whale stopped diving deep and spent the next five hours near the surface, moving slowly. Looking at the dive data, it was obvious what was going on, and this is a pattern we've seen many times since. The whale was following its prey, primarily squid, and foraging throughout the night. The prey of short-finned pilot whales in Hawaiʻi appear to associate with a dense layer of small marine organisms called the "deep-scattering layer," as it reflects the signals of echo sounders deep in the water column. Organisms in this layer generally avoid light, so they stay deep during the day and rise closer to the surface around sunset. The deepest dives were during the day when the prey were down deep, and as these organisms rose toward the surface the pilot whale dives became shallower. This intensive diving all night long reflected a long period of feeding, and by 4 a.m. the whale was probably full. Shortly after, the prey descended to avoid the morning light, and the whale rested.

Over the last fifteen years, short-finned pilot whales have been our most frequently encountered species of odontocete around the main Hawaiian Islands, and we've tried to use these opportunities productively. Through the end of 2015, we had encountered them 605 times, taken more than a quarter million photos, collected 250 biopsy samples for genetic and toxicology studies, and deployed more than one hundred satellite tags on individuals off all the islands. Although

A group of short-finned pilot whales off Kona, April 17, 2015. The individual on the far right is an adult female, HIGm0605, first documented off Kona in 2005. Photo by author.

we know a tremendous amount about this species in Hawaiian waters, there is still a lot to learn.

Identifying Features and Similar Species

There are two different forms of short-finned pilot whales recognized in the North Pacific, noted off Japan as early as 1760 as *shiho goto* (*goto* means "pilot whale"), found off northern Japan, and *naisa goto,* found off southern Japan. Although they were once thought to be different species, the two forms were later determined to be the same species. Recent research shows that they are likely different subspecies, as they are genetically distinct and have different habitats throughout

An adult male short-finned pilot whale off Kona spyhopping, May 6, 2012, showing the very squared-off head typical of *naisa goto*. Lighter blazes are also visible, extending from above the eye toward the dorsal fin. Photo by Jessica M. Aschettino.

Short-finned pilot whale adult male (top), adult female (middle), and juvenile (bottom).
Illustrations by Uko Gorter.

the North Pacific. *Shiho goto* are larger, and have a fairly distinct lighter-colored saddle patch behind the dorsal fin (similar to killer whales, but wispier and farther back), and adult males have a rounded head when viewed from above. Adult males of *naisa goto* have a squared-off head when viewed from above, and the saddle patch is indistinct. The individuals in Hawaiian waters all appear to be the smaller *naisa goto* form like the ones off southern Japan, while pilot whales off the west coast of North and Central America are the larger *shiho goto* form.

Short-finned pilot whales are the second-largest delphinid in Hawaiian waters. They are born at about 1.4 m (~4′7″), and the largest individual measured

An adult male short-finned pilot whale surfacing off Kona, May 6, 2012, showing the broad base of the dorsal fin and the proximity of the fin to the blowhole. This individual, HIGm0460, was first documented off Kona in 2005. Photo by Daniel L. Webster.

in Hawai'i, an adult male that stranded on Lāna'i in 2013, was 5.5 m (~18′), possibly the maximum size noted for the smaller *naisa goto* form. They are dark gray in color, often appearing all black, with a very faint gray blaze between the flippers. In good lighting conditions, a faint dorsal cape can be seen, but the contrast between the cape and the lateral coloration is much less than for other Hawaiian blackfish. The saddle patch cannot typically be seen except in underwater photos or photos in extremely good lighting conditions. While pilot whales in Hawai'i are bitten by cookie-cutter sharks, the scars heal to the same color as the background, so they are not particularly noticeable.

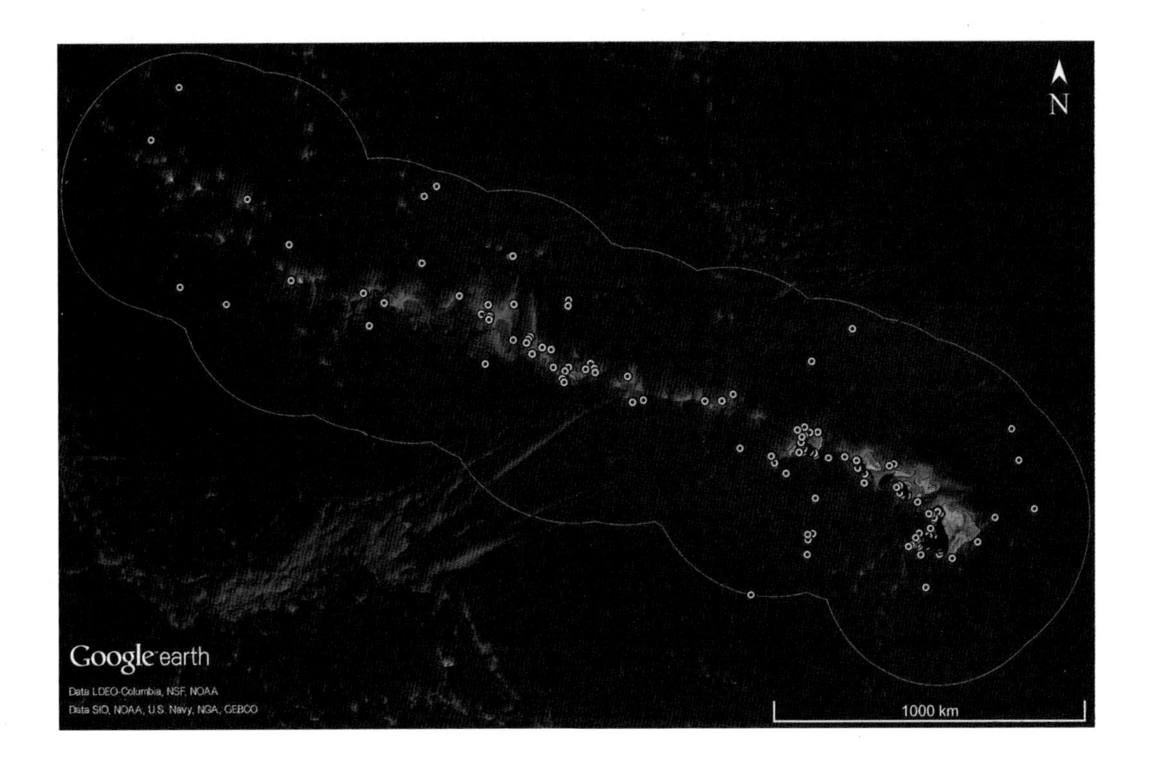

Short-finned pilot whale sighting records in Hawaiian waters. While sightings are concentrated around the islands, they are also regularly documented in offshore waters.

The most distinctive feature of the short-finned pilot whale is the broad base of the dorsal fin—for adults of both sexes, the base is more than twice as long as it is tall. Unlike any other species in Hawaiian waters, the fin is located forward of the center point of the body, so the distance from the front of the fin to the blowhole is relatively short. The relative size and shape of the dorsal fin varies between the sexes and with age. For older adult males, the distance from the front of the fin to the front of the head is only a little bit longer than the length of the fin's base. They have a rounded head that becomes progressively flatter on the front in older males (and to some degree in older females). Adults also differ dramatically in body size, with adult males reaching lengths of about 1.09 m (3′7″) longer than adult females on average. Females continue to grow until they are about twenty-two years of age, while males continue to grow until about twenty-seven years of age.

Short-finned pilot whales are confused most often with false killer whales, but the longer and relatively short dorsal fin, placed farther forward on the body, is the easiest way to distinguish a short-finned pilot whale from false killer whales or the smaller blackfish—pygmy killer whales and melon-headed whales. Short-finned pilot whales also spend a lot of time during the day logging (resting) at the surface, a behavior rarely observed for false killer whales.

Habitat Use, Movements, and Abundance

Short-finned pilot whales are found in tropical and subtropical waters worldwide, both in the open ocean and on continental and island slopes. In Hawaiian waters they are found in the open ocean, around the Northwestern Hawaiian Islands, and around the main Hawaiian Islands, but density is higher around the main Hawaiian Islands. In our surveys, we've recorded pilot whales in water as shallow as 324 m and as deep as 4,400 m, but there is a clear peak in sighting rates between depths of about 500 and 3,000 m.

We know a lot about their movements, both by looking at resightings of individuals among the islands and by satellite tagging, and we can use the resighting history of tagged individuals to help understand and interpret the tag data. As of the end of 2015, we had over 1,200 distinctive individuals in our photo-identification catalog, and we had data from over one hundred satellite tag deployments. While more than half the tags were deployed off Hawai'i Island, we have tagged individuals off Kaua'i, O'ahu, and Lāna'i as well. These 1,200 individuals likely represent about a hundred different social units or "pods." From resightings of distinctive individuals, we know that pilot whales are long-term residents around the main Hawaiian Islands; individuals Dan McSweeney documented off Kona in the late 1980s are still around today. From analyses of association patterns of individuals identified among all the islands undertaken by Sabre Mahaffy, the curator of most of our photo-identification catalogs, there appear to be several different communities of resident short-finned pilot whales, perhaps similar to the communities of "resident" killer whales in the Pacific Northwest. These communities seem to be socially isolated, and each has a home range that centers around one or two island areas but that overlaps with the home ranges of other communities.

Off the island of Hawai'i, almost all the satellite-tagged individuals have remained around the slopes of the island, with some moving to the slope off the north side of Maui. Movements of individuals from the Hawai'i Island community offshore or to other areas among the islands are extremely infrequent. The core range for this community extends from near Ka Lae (South Point) up the west side of the island to the area off of 'Upolo, the northern tip of the island. Interestingly, there is an area off the east side of Hawai'i Island that seems to be avoided, roughly from the easternmost point, Cape Kumukahi, to just north of Hilo. We haven't tagged any whales off the Hilo side of the island, however, so it is possible that there are just different groups in that area that rarely make it to Kona.

There is also a community whose home range is centered around Kaua'i and Ni'ihau but that extends offshore and to western O'ahu. These individuals use deeper water much more often than those from the Hawai'i Island community, probably because they have no choice; the smaller sizes of the islands make for insufficient slope habitat to sustain the community, so they range in offshore waters more often. In between, there is another community that we know less about that seems to extend generally from O'ahu to Lāna'i and Moloka'i and overlaps with both of the communities to the east and west. Relatively little work has been done with this community, so more is yet to be learned, both about them and about individuals from the pelagic population.

We have tagged individual pilot whales in several groups not associated with any of the island communities. Most of these have not been seen before or since, and based on their movement patterns they were clearly individuals from an open-ocean population. The individuals have meandered over a wide area offshore of the islands, have moved through the deep-water channel between Kaua'i and O'ahu, and have moved into international waters. We've never documented these apparently offshore animals socializing with individuals from the resident island communities, although it is certainly possible they do so on occasion.

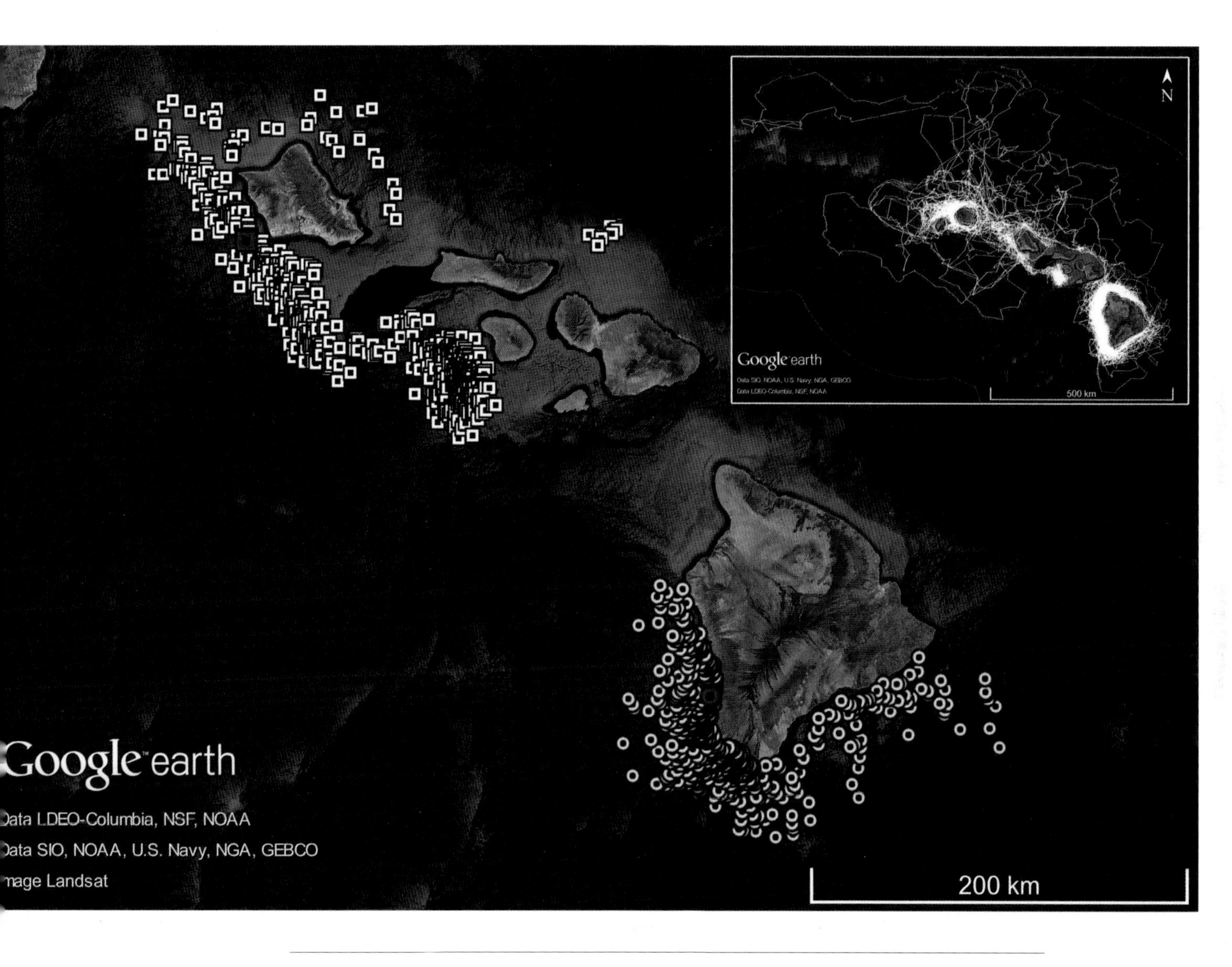

Some short-finned pilot whales are homebodies. This map shows satellite tag data from two individuals from island-associated communities, one tagged off O'ahu and tracked over 229 days (white squares) and another tagged off Hawai'i Island and tracked over 110 days (yellow circles), both staying close to the islands where they were tagged. Tagging locations are shown with red symbols. The inset shows satellite tag data from seventy-one individual pilot whales tagged since 2006. Most offshore movements are of animals thought to be from an open-ocean population.

Abundance has not been estimated for the island-associated population or any of the specific island communities, but in all of Hawaiian waters there are an estimated nineteen to twenty thousand short-finned pilot whales. Around the main Hawaiian Islands, as noted earlier, they are the most frequently encountered species in our work, which reflects both their overall abundance and how easy they are to detect, given their group sizes, size of individuals (particularly adult males), and frequent behavior during the day of logging at the surface. On an island-by-island basis, it is only off Hawai'i Island that they are the most frequently encountered species, while off the other islands they are the fourth most frequently encountered species in our work.

Predators and Prey

Although it is possible that large sharks may occasionally attack short-finned pilot whales, surprisingly we have no photos of individuals that have obviously survived an attack from a large shark. Given that we see individuals of other smaller species with such scarring, it suggests that large sharks tend not to attack pilot whales, or the attacks are only on small individuals and are almost always fatal. There is one individual from the Hawai'i Island community that has scars from an interaction with a large toothed whale; from the spacing of the tooth rakes, it was almost certainly from a killer whale attack.

We have a good idea where and when short-finned pilot whales are feeding, but we don't know much about what they are actually feeding on. When they feed during the day it is typically at great depths, usually between 700 and 1,000 m; the deepest we've documented pilot whales diving during the day was 1,552 m. These deep dives typically last about fifteen to twenty minutes, but we've documented one adult male diving as long as twenty-seven minutes. At night, particularly around sunset, they also dive very deep, often over 1,000 m, but during most of the night the foraging dives are typically from 300 to 500 m. Occasionally they bring their prey up to the surface and often either discard it or accidentally lose it. We pick up any squid or fish we see floating at the surface that might be prey of pilot whales or other deep divers. Around pilot whales, we've found one deepwater eel and a number of squid. Elsewhere, short-finned pilot whales typically feed on a diversity of midsized squid and small deepwater fish.

Life History and Behavior

Large numbers of short-finned pilot whales have been killed in whaling operations off Japan, and as a result considerable information is known about their life history. Females become sexually mature at an average of about nine years of age and have their first calf at ten or eleven years of age. In Hawaiian waters, calving seems to occur between July and November, with a peak in July. Calving interval is about five years, and an average female may have four to five calves over her lifetime. Like false killer whales, killer whales, and humans, females go through menopause. Female short-finned pilot whales stop reproducing in their middle to late thirties, but they may live for thirty or more years afterward. These older females may be the leaders of the social units or "pods" described earlier. Males mature later, hitting puberty in their midteens. Mortality of newborns is probably quite high, but then it decreases until about twenty-eight years of age for males and about forty-six years of age for females. Females may live to their midsixties or older, while males seem to have a harder life, with most living only to their midforties.

While we have seen lone short-finned pilot whales on a few occasions, they have all been adult males. In general, pilot whales are a social species, with an average group size of eighteen around the main Hawaiian Islands. Twice we've seen large aggregations, two different groups estimated at 185 and 195 individuals respectively, seen a week apart off Kona in 2009; these were probably temporary aggregations of a number of smaller, more stable social units. Groups of more than fifty individuals are quite rare, representing less than 1 percent of our sightings. Smaller groups are likely relatively stable social units—individuals do have preferred companions—and these stable groups range in size from about five to sixteen individuals. Not all individuals stay in their natal group, however, as is seen in fish-eating "resident" killer whales and possibly false killer whales in the main Hawaiian Islands. We occasionally see groups that appear to be composed entirely of adult males, subadult males, and juveniles that we suspect are males. It appears that most males may disperse from their natal group, possibly in their early teens.

In about 10 percent of our encounters with short-finned pilot whales, they are seen with other species of whales or dolphins, most often humpback whales but also pantropical spotted dolphins or rough-toothed dolphins. Other species

Like many other animals, pilot whales grieve for their dead. These three individuals are from the Hawai'i Island community and were photographed carrying a dead calf on July 12, 2015. The adult male (HIGm0511) carrying the calf and the female in the top of the photo (HIGm0516) were first documented as adults in 2005. The individual on the bottom, HIGm0905, was first seen in 2007 and was a juvenile at that time, and she is probably the offspring of HIGm0516. The mammary glands of HIGm0905 are distended, suggesting she may be the mother of the dead calf and had not nursed for several days. Photo by Deron S. Verbeck/iamaquatic.com.

seen with them include Cuvier's beaked whales, melon-headed whales, pygmy killer whales, Risso's dolphins, false killer whales, bottlenose dolphins, and even a fin whale. With larger species such as humpback whales and the fin whale, the pilot whales appear to be harassing them, while the roles are often reversed with smaller species, where the pilot whales are victims of harassment. We once witnessed a large group of melon-headed whales harassing a group of pilot whales. The response of the pilot whales was to group up and all dive together, diving deep enough so that the smaller melon-headed whales could not follow them, and then leaving the area entirely while down on their long dive.

The most common interspecific association we see with pilot whales is actually with oceanic white-tip sharks. These sharks often appear to follow pilot whales around, presumably to scavenge lost or discarded prey, or possibly even eating whale feces. It is also likely that the sharks take advantage of the whale's superior abilities to find concentrations of prey using echolocation and follow the whales down on deep dives. We actually tried to test this back in 2002, deploying a depth-transmitting tag on an oceanic white-tip shark at the same time as we had several whales tagged with time-depth recorders, but the shark left the group of whales shortly after we tagged it.

Short-finned pilot whales, along with their cold-water sibling species the long-finned pilot whale, are more prone to mass stranding than perhaps any species, probably in part because of their strong social bonds. There have been five mass strandings of pilot whales reported in Hawai'i, occurring on Kaua'i, O'ahu, Moloka'i, and Lāna'i. As noted earlier, four of these were between 1957 and 1959 and involved a total of sixty-six individuals, most of which died. The most recent was a group of four animals reported from Moloka'i in 1989. Given the tendency of this species to mass strand elsewhere and these records from Hawai'i, they are probably the most likely species to mass strand on Hawaiian shores in the future.

Short-finned pilot whales spend much of their day resting at the surface, often logging. When approached by boats, they often spyhop—rising vertically out of the water headfirst to look around and at the approaching boat. As long as boats don't approach too quickly or too closely, pilot whales are usually indifferent to them, but this behavior seems to vary among the islands, with pilot whales off Lāna'i avoiding boats more often than those elsewhere, perhaps because there is so little vessel traffic to the west of Lāna'i and they are not used

Oceanic whitetip sharks often associate with pilot whales in Hawai'i. They likely scavenge on scraps left behind by the whales and feed on whale feces, but it is also possible they follow the whales down to find prey at depth, taking advantage of the whales' echolocation abilities. This photo was taken off Kona, June 19, 2015. Photo by Deron S. Verbeck/iamaquatic.com.

to their presence. Occasionally, individuals will approach boats; often these are subadult or adult males that are either curious or perhaps aggressive, charging the bow of the boat and turning away at the last second, sometimes throwing water up onto the bow. They usually either ignore swimmers or dive to avoid them, although there are always exceptions.

In a well-known 1992 incident shown dozens of times on television, Lee Tepley, a photographer, and Lisa Costello went in the water off Kona with a group of pilot whales that had been engaged in high-speed swimming and intensive social interactions. Lisa and an adult male approached each other slowly, and Lisa began petting the whale. Shortly after she stopped, the whale turned and grabbed Lisa by the thigh, let her go, and then apparently grabbed her again, taking her down at least 10 to 12 m below the surface before bringing her back up to the surface. In a paper published later in the journal *Marine Mammal Science,* Shane, Tepley, and Costello note that this "near-death experience . . . should serve as a reminder to humans that cetaceans, even when behaving non-aggressively, can pose a serious threat."

In 2013, a similarly dangerous interaction occurred with a pilot whale off O'ahu. Three free divers wearing snorkeling gear got in the water several times near a group of short-finned pilot whales. One individual, a juvenile based on photos, approached the free divers and interacted with them. The whale mouthed the swim fin of one person and ended up pulling it off, taking it down and shaking it side to side, like a dog with a toy. At that stage, the free divers got out of the water, but two ended up going back in a few minutes later. The same whale rapidly approached and grabbed the heel of one person, pulling him down, but he was able to get his foot out of his fin and make it back to the surface and into the boat. The whale then diverted its attention to the remaining free diver in the water, approaching her rapidly with its mouth open. Wearing a monofin (a single fin for both feet), she used it to fend the whale off, pulled herself over the top of the whale toward the boat, and as she was entering the boat the whale bit her once in the leg, puncturing her wet suit and bruising but not breaking the skin. Why this interaction occurred is unclear—a young whale, perhaps a male, got curious and rambunctious with people in the water. Although these incidents suggest caution is warranted, such events are extremely rare given the number of times swimmers have entered the water near pilot whales.

Conservation

Like most other species of whales and dolphins in Hawaiian waters, short-finned pilot whales are at least occasionally affected by human activities. When proposed regulations were first mentioned to limit interactions between swimmers and spinner dolphins, we saw an increase in the number of tour vessels swinging offshore of Kona, interacting with and often putting swimmers in the water with short-finned pilot whales. What affect this has on their resting patterns is unknown, but there is certainly a potential there for harassment.

Two adult male short-finned pilot whales off Ni'ihau, February 12, 2015. A satellite tag is visible on the dorsal fin of the animal to the left. These whales are from the pelagic population. None of the twenty distinctive individuals in this group had been previously documented, and over the forty-five days tag data were obtained they roamed widely in deep waters offshore both to the west and east of the islands. Photo by author.

Because of their tendency to rest motionless at the surface, they are sometimes hit by vessels, and there are a few individuals from the island-associated population that have survived with propeller wounds visible. We've also seen individuals trailing fishing lines out of their mouths, suggesting that at least on occasion they try to take the bait or catch from fishermen and are hooked as a result. If they ingest the hook or have a large amount of trailing gear, this could easily lead to death. Short-finned pilot whales are also occasionally hooked and killed or seriously injured in the offshore longline fishery for *'ahi po'onui,* the second most frequently recorded species of whale or dolphin caught in that fishery.

Another conservation concern is the impact of high-intensity sounds—in particular, high-intensity mid-frequency active (MFA) sonar used by the U.S. Navy and other navies in Hawaiian waters. Off North Carolina in 2005, there was a multispecies stranding that included thirty-three short-finned pilot whales at the same time that the navy was using MFA sonar offshore. During the 2014 Rim of the Pacific Exercise (RIMPAC), a short-finned pilot whale live stranded and died in Hanalei Bay, Kaua'i, another individual was seen in very shallow water in the bay a few days later, and a third individual live stranded and died on O'ahu. Whether these strandings were caused by sonar used during RIMPAC may never be known, but it is certainly possible that naïve animals in particular may be seriously impacted by MFA sonar, and we do encounter groups in the area that appear to be from the pelagic population.

Interestingly, the core range of the Kaua'i and Ni'ihau community of short-finned pilot whales overlaps with an area used regularly by the U.S. Navy for training and testing purposes. Members of that community have likely been exposed to MFA sonar repeatedly throughout their lives over the last forty years. We've had satellite tags on some of these community members before and during sonar use in the area, and we have not seen large-scale movements out of the area. This suggests that at least some species may tolerate the high-intensity sounds to some extent, particularly if the area is an important one to them.

PYGMY KILLER WHALES *(Feresa attenuata)*

When the first pygmy killer whales were discovered in the wild in the 1950s and early 1960s, things did not go well for them. The first live individual was harpooned off Taiji, Japan, in 1952 and brought in for processing. Although the animal was quickly cut up, the parts were retained and a local researcher was able to patch them together into a Frankensteinian reconstruction that enabled him to describe the external appearance for the first time. When he compared the skull to others, it was a species that had originally been discovered based only on two skulls, one found in 1827 and the other in 1874. Based on similarities in the size and number of teeth, as well as the tall dorsal fin and white markings in the genital area, he proposed "lesser" or "pygmy killer whale" as the common name, and the latter seems to have stuck. As it turns out, they are in fact related to killer whales, but their closest relatives are the other blackfish: false killer whales, pilot whales, and melon-headed whales.

The next two times individuals of this species were knowingly encountered in the wild, their lives were similarly cut short by humans. The first was an individual killed off Senegal, Africa, in 1958. A group of fourteen encountered off Japan in 1963 was taken into captivity, where they all died within twenty-two days. As Karen Pryor noted in her book *Lads before the Wind,* on July 6, 1963, the collector for Sea Life Park radioed the park from Kona to let them know that he "saw a school of really weird porpoises . . . black all over except for white lips; he called them 'clown' porpoises." Ten days later he encountered the group again and captured one of the individuals. When they brought it on board, it snapped at its captors, and when first introduced into the tank at Sea Life Park, Karen said "I was its nearest target, and it leaped for my face, teeth clashing, like a seagoing wolf." Such behaviors are rarely seen by any species of whale or dolphin, in captivity or in the wild, and clearly pygmy killer whales are in a class by themselves.

Since the 1960s, pygmy killer whales have been documented in a number of tropical areas worldwide, but they remain a poorly known species except in Hawai'i. Dan McSweeney originally began taking photos of these whales off Kona in the 1980s, and while there are still major gaps in our knowledge of this species, the Hawaiian population is now the best known anywhere in the world.

Pygmy killer whales off Kona. The individual in the middle, HIFa003, was first documented off Kona in 1991. The individual on the right, HIFa012, was first seen in 1996 and is an adult male, with a distinct post-anal ventral keel. This photo was taken in 2009. Photo by Deron S. Verbeck/iamaquatic.com.

Identifying Features and Similar Species

Pygmy killer whales are about 80 cm (2′7″) at birth and reach maximum lengths of 2.6 m (~8′6″). The front of the body is quite robust and tapers toward the tail. They have a rounded head and, in poor lighting conditions, appear largely black in color. In good lighting it is possible to see the dark-gray cape and lighter-gray sides, with a very distinct boundary between the two. On the crown of the head, the cape broadens to a darker, larger patch in front of the blowhole to the front of the melon. Although they don't show it often, they have a large white patch on the belly and a lighter-gray blaze sometimes visible between the flippers, and older animals typically have numerous white oval scars on the chest and belly from cookie-cutter shark bites.

This adult female pygmy killer whale, HIFa005, was first documented off Kona in September 1992. The photo was taken December 6, 2008, but she was seen as recently as November 2015. It shows the clear demarcation of the dorsal cape just in front of and below the dorsal fin, as well as the darker patch on the crown. Photo by author.

Pygmy killer whales have large, widely spaced teeth, and most adults tend to have one or more sets of white tooth rake scars visible on the body or head. Older animals not only have white lips but also white that extends onto the chin and front of the head. Males and females look more or less the same, but adult males have a distinct bulge, called a post-anal hump or a ventral keel (like the keel of a sailboat), which can be easily seen underwater. Pygmy killer whales can be quite difficult to spot, as they spend a lot of time during the day resting motionless at the surface like dwarf sperm whales, often with just the back visible and the dorsal fin underwater.

A pygmy killer whale lunging at the surface during social activities off Kona, November 6, 2011. This individual, HIFa249 in our photo-identification catalog, is an adult male first documented in 2005. Extensive white pigmentation on the front of the face and the white scars on the belly from cookie-cutter shark bites are visible. Photo by author.

Pygmy killer whales are most frequently confused with melon-headed whales, but there are a few characteristics that can be used to tell them apart. The clearly demarcated dorsal cape projects downward at a shallow angle in front of and below the dorsal fin, compared to the diffuse and steeply projecting cape of the melon-headed whale. The darker cap on the crown of the head is also diagnostic: melon-headed whales have only a dark facial mask. Pygmy killer whales often have paired white linear scars—tooth rakes caused by other pygmy killer whales—while such scars are rare on melon-headed whales. Pygmy killer whales have rounded tips on their flippers, while melon-headed whales have pointed flippers, although getting a good view of the flippers is difficult. Pygmy killer whales might also be confused for small false killer whales; the most obvious

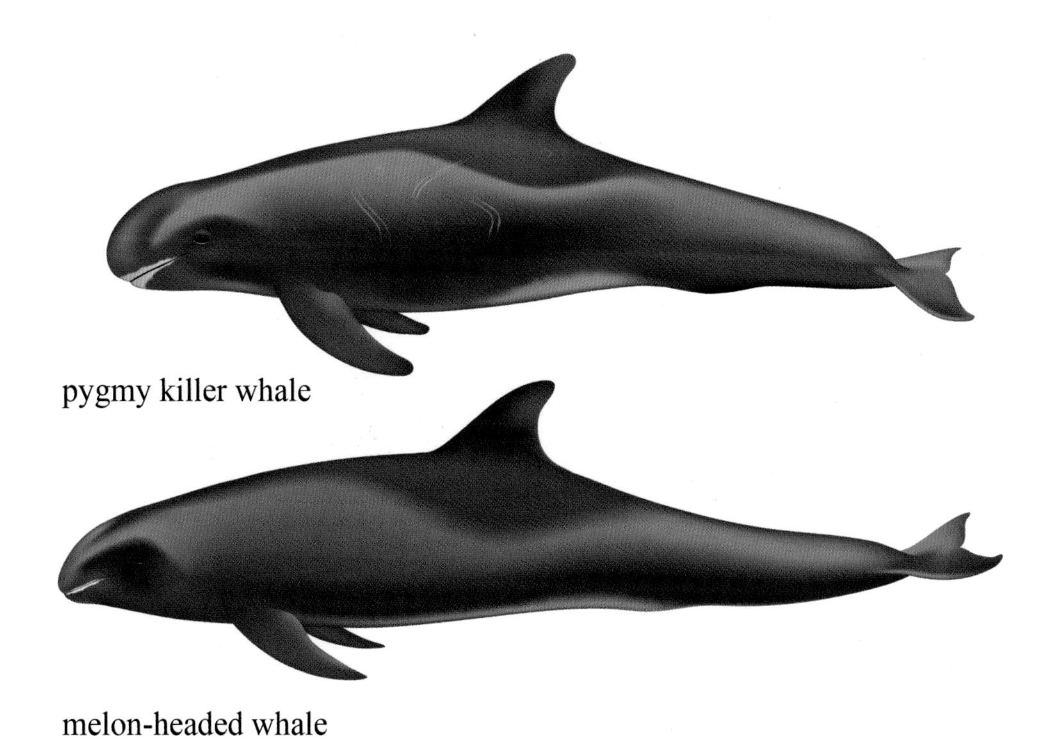

A pygmy killer whale (top) and a melon-headed whale (bottom). Illustrations by Uko Gorter.

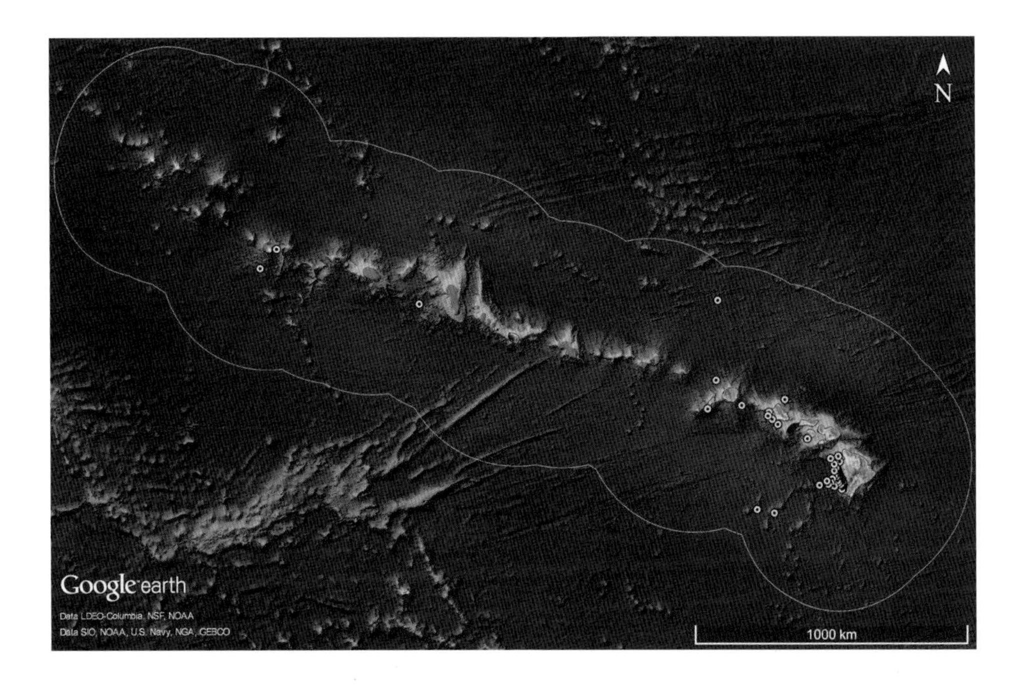

Pygmy killer whales are seen infrequently in offshore Hawaiian waters, with most sightings around the main Hawaiian Islands off O'ahu and Hawai'i Island.

difference is the size of the dorsal fin, which is larger in proportion to body size in pygmy killer whales than in false killer whales. Pygmy killer whales tend to move much slower than false killer whales and rarely engage in aerial behavior, and it would be unlikely to find more than one or two juvenile false killer whales together without having an obviously larger adult present.

Habitat Use, Movements, and Abundance

Pygmy killer whales are found in tropical open-ocean waters worldwide. The only known exception to this is in Hawaiian waters, where they are resident on the island slopes, at least off of Oʻahu, Maui Nui (the collective name for Maui, Molokaʻi, Lānaʻi, and Kahoʻolawe), and Hawaiʻi Island. They have been documented on a few occasions in offshore waters and around the Northwestern Hawaiian Islands, but the vast majority of sightings come from the main Hawaiian Islands, where they are most commonly seen between depths of 500 and 3,500 m. Our shallowest sighting was in 114 m, but it was off Kona, where the slope is quite steep. We've encountered them off Ni'ihau, Kaua'i, O'ahu, Lāna'i, and Hawai'i

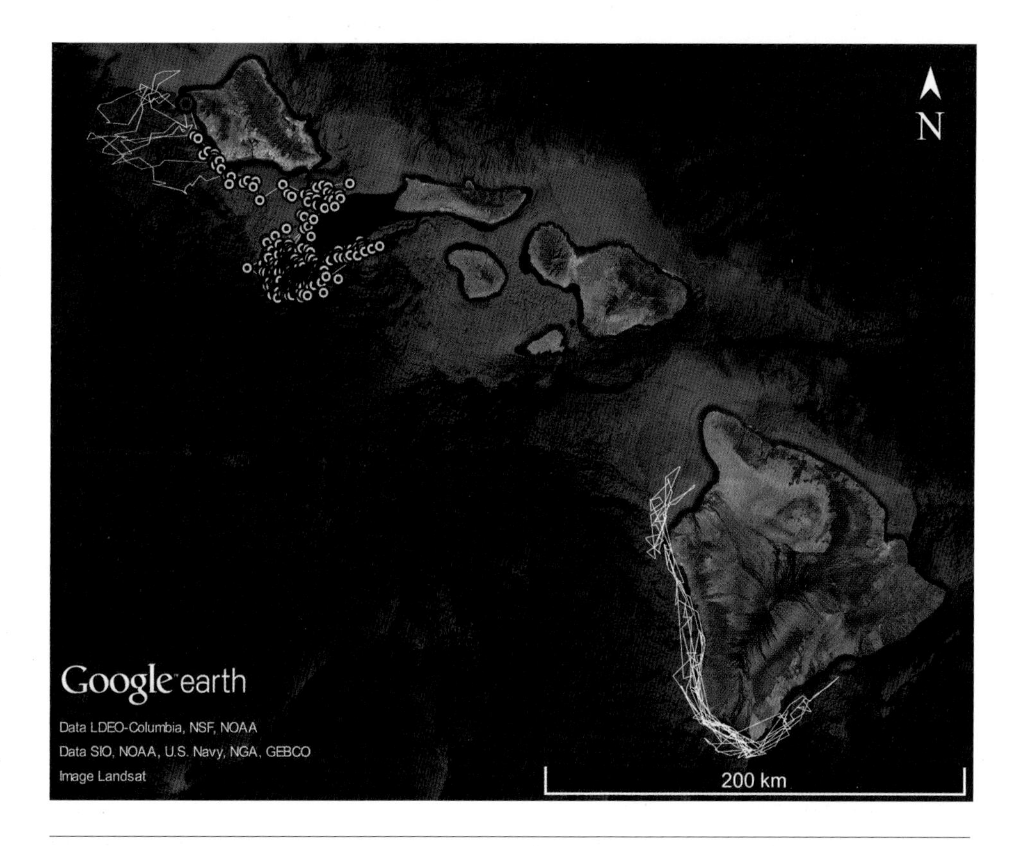

Satellite tag data from five pygmy killer whales, three tagged off O'ahu in 2010 and two tagged off Hawai'i Island, one in 2008 and one in 2009. Locations (yellow circles) are shown for two individuals tagged within the same group over a thirty-day period, with the initial tagging location shown with a red symbol. This group has been regularly seen off O'ahu since 2007 and appears to be resident to O'ahu and Penguin Bank.

Island, although they appear to be less common off Ni'ihau and Kaua'i than around other islands, and they rarely enter the Maui Nui basin.

On January 1, 1986, Dan McSweeney photographed a group of eight distinctive pygmy killer whales off of Kona. In succeeding years, all eight were seen off the island again, and one of them, an adult-sized individual at the time, was seen in the same area as recently as May 2013, more than twenty-seven years after first being documented. This individual, affectionately referred to as HIFa001 (the first *Feresa attenuata* in our catalog), was seen forty times off the island in the intervening years. This isn't the only individual resighted—another individual first seen with HIFa001 in 1988 has been seen thirty-five times over the years, most recently in 2013, still associated with HIFa001. Over the last thirty years, more than two hundred distinctive individuals have been documented off Hawai'i Island, and almost half have been seen on more than one occasion, with most of those encountered in multiple years. Clearly, there is a community of pygmy killer whales that is resident to the waters off the island of Hawai'i.

In 2007 we began to get photos of pygmy killer whales off O'ahu taken by Tori Cullins of the Wild Dolphin Foundation, and many of those individuals have been seen off O'ahu every year since, evidence of a resident community there as well. We have documented interisland movements based on photo identification, but such movements are rare. Some of the O'ahu residents were seen off Kona in 2012, although they were not apparently associated with any of the Kona residents, and they have since moved back to O'ahu.

We've satellite tagged individuals from only three different social groups, and all have spent most of their time on the leeward sides of the islands. Two different groups satellite tagged off O'ahu showed quite different movement patterns. We tagged two individuals from the O'ahu resident group, and they stayed in relatively shallow water less than 1,000 m deep, but they spent a lot of time on the edge of Penguin Bank. The other group had been seen off O'ahu only once before, as well as once each off Lāna'i and Hawai'i, and spent their time moving about in depths from 1,000 to 3,000 m deep. These types of differences likely reflect that one group was in the core of their range, hunting in the best areas that were very familiar to them, while the other was a group visiting the area and feeding on different resources farther offshore.

Pygmy killer whales are naturally rare. In the eastern tropical Pacific, they ranked twelfth out of thirteen species of delphinids that had abundance estimates. In our work, they are the least frequently encountered of the eleven resident species, representing less than 2 percent of our sightings of odontocetes. This reflects both the difficulty in spotting pygmy killer whales and their rarity. There are two estimates of abundance for Hawaiian waters: 956 individuals from the 2002 NMFS survey and over 10,000 individuals from the 2010 NMFS survey. There is a lot of uncertainty associated with both estimates, as well as differences in methods between the two surveys, and I suspect that the true abundance of this species in Hawaiian waters is somewhere in between.

Predators and Prey

Like other species of odontocetes, pygmy killer whales are no doubt occasionally attacked and killed by killer whales in Hawaiian waters, but it is large sharks that are probably their most important predator. One juvenile individual seen off Kona in 2008 appeared to have been repeatedly mouthed by a large shark, as it had numerous scars on the dorsal surface. Like melon-headed whales and dwarf sperm whales, many of the individuals with apparent bite wounds from large sharks have been bitten on or around the dorsal fin. Are these species more vulnerable to attacks because they spend so much of their time during the day resting motionless at the surface? Or are they more likely to survive attacks because they roll over at the last second before being bitten, so that the bites occur on the less vulnerable dorsal surface? I suspect both of these may be true.

Pygmy killer whales in Hawaiʻi seem to do most or all of their feeding at night. During our daytime surveys we've seen them forty-eight times, and Dan McSweeney has seen them about sixty-five times, and in none of those encounters has there been any evidence of feeding. What they feed on in Hawaiʻi is unclear—the stomachs of a couple of animals that have stranded have been empty. Elsewhere they have been reported feeding on squid and small fish, and with their relatively large teeth it wouldn't be surprising if they feed on larger squid and fish. There is some suggestion that pygmy killer whales might feed on other smaller dolphins, although we've seen no evidence of it in Hawaiʻi. Like false killer whales, pygmy killer whales have been implicated in the killing of dolphins that were released,

A pygmy killer whale calf off Kona, December 9, 2008, bearing the scars of a shark attack. The apparent mother of this individual, HIFa009 in our catalog, has been seen sixteen times since December 2008, as recently as December 2014, but the calf has not been resighted. Photo by author.

often disoriented and injured, from purse seine nets in the eastern tropical Pacific. In captivity, unlike most other species of whales or dolphins, pygmy killer whales have aggressively attacked other species. An individual captured off Kona in 1965 later killed a young pilot whale in the same tank, and one that stranded in South Africa killed a young dusky dolphin it was held with, but in neither case did they show any signs of trying to feed on the victim.

Life History and Behavior

Assessing the life history of a species is usually done with dead animals, either killed deliberately or taken from mass strandings, and then aging individuals by looking at layers in the teeth and examining the gonads for evidence of pregnancy or sperm. Pygmy killer whales have rarely been hunted, and mass strandings are infrequent given their rarity, and thus not a lot is known about their life history. From observations and from similar species, we can infer some aspects of their life history. They are probably slow to mature and have long calving intervals, as do closely related species such as false killer whales and pilot whales—we rarely see newborns or yearlings. The few newborn animals that have been documented in Hawaiian waters were only in August and October, suggesting that they may have more seasonal reproduction than many species of odontocetes in Hawai'i.

Group sizes of pygmy killer whales in Hawai'i have ranged from pairs of individuals to one group of thirty-three, but the average group size is nine individuals. Usually individuals in groups are fairly close together, typically spread out at most over a couple of hundred meters. The only widely dispersed group we've seen was also the largest, with a total of thirty-three individuals spread out over more than a kilometer, but this was likely a temporary aggregation of several more stable groups together. From analyses of associations, pygmy killer whales have extremely strong and enduring social bonds. Two adult females first seen together in 1994 were still together twenty years later, and in almost every sighting of either individual in the intervening years, both were present. One other stable group we've documented off Kona has four adult males and one adult female, and most of these individuals have been seen together over more than ten years. Whether these stable groups are closely related individuals or just long-term companions is not known.

[Top] During the day, pygmy killer whales spend much of their time resting at the surface and often spyhop when boats approach. This photo was taken off Kona, December 6, 2008. Photo by Daniel L. Webster. [Bottom] Pygmy killer whales are often somewhat elusive but seem most interested in boats that are motionless in the water. This individual approached our research boat off O'ahu, October 13, 2010, and appeared quite curious about us. Photo by author.

Pygmy killer whales seem to spend most of the day resting, slowly traveling, and socializing; as noted, we've never seen them feeding, and fast travel is also rare. They are occasionally seen with other species, most often short-finned pilot whales, and sometimes they are mixed in with the pilot whale group. We've also seen them near bottlenose dolphins and false killer whales, but they have been just passing each other by. Their reactions to boats are extremely variable and likely depend on what they were doing when they were interrupted. Sometimes they show a lot of curiosity about boats and apparently the people in them, circling on their sides clearly looking up into the boat, and sometimes spyhopping, facing toward a boat. They seem more likely to do this when they are already engaged in social activities. When we have approached a resting group, they sometimes avoid the boat, although they tend to do it in a slow and measured fashion, turning away from the boat or slowly diving as it approaches. They don't act as if they are frightened, like striped dolphins or Fraser's dolphins, but rather they just have a large personal space and don't want anyone or anything to invade it. I've found that groups that are otherwise disinterested in being approached often become curious when I put the boat into neutral 50 or 100 m away from them. After a while I suspect they start to wonder why we aren't approaching and may approach us instead. Some of our best opportunities for tagging or photography have occurred when I've put the boat into neutral near a group.

Conservation

Starting in 1968 a number of mass strandings and near mass stranding events have been documented for this species. Given their rarity and the evidence of small resident populations as in Hawai'i, such events have the potential to impact an entire population. Mass strandings have been recorded in a number of locations around the world, particularly off Taiwan and on the coasts of Florida and Georgia, but they've also been recorded in Hawai'i. In 1981, four individuals stranded at Mā'alaea, Maui; one died, but three were returned to the water, although their fate is unknown. In 1988 a similar event happened in the same area: at least six pygmy killer whales were seen swimming in very shallow water, two of which stranded and died, and the rest were not seen again. The Mā'alaea area does not seem to be a good area for this species. In 2009 a group of at least seven pygmy killer whales spent several weeks in very shallow water in this area, and at least one of the group died. In that case, none of the individuals involved had been seen before or have been since, but it is hard to say whether they were from an open-ocean population or an island-associated group that perhaps primarily occupies the windward sides of the islands, where we have few photos.

There are probably many reasons why such strandings occur. With such strong social bonds, if one individual in a group is sick, all members of the group might move into shallow water to support it during its last days, placing themselves at risk. Two mass strandings of pygmy killer whales, one off New Caledonia and one off the Virgin Islands, were associated with major hurricanes, and those animals may have been pushed into shore or moved close to shore to shelter from the storm. Since 1996 four mass strandings and three near mass stranding events have been documented off Taiwan, including the two largest groups of pygmy killer whales stranding anywhere: eighteen and twenty-eight individuals. This is an area where naval exercises from a number of countries have been undertaken, and researchers have suggested that the increase in mass strandings of pygmy killer whales in that area is due to their susceptibility to impacts from naval sonars. It is interesting to note that while there are apparently resident populations of pygmy killer whales off Hawai'i Island, O'ahu, and Penguin Bank, sightings are extremely infrequent off Kaua'i and Ni'ihau, where most of the naval activity in Hawai'i occurs.

Pygmy killer whales also face other risks of impacts from human activities. One pygmy killer whale was recently hooked and died in the offshore longline fishery. Because they don't seem to feed during the day, they haven't been observed taking bait or fish off fishermen's hooks around the main Hawaiian Islands, but one individual pygmy killer whale that stranded on O'ahu in 2006 had a healed scar both inside the mouth and along the lip, indicating that it had previously been hooked in some sort of fishery. Pygmy killer whales off Hawai'i Island and O'ahu frequently have cuts and other types of injuries along their mouths that show they must be taking bait or fish from hooks on a pretty regular basis and are hooked at least some of the time. How often such hooking ends up killing animals is unclear, as such a small proportion of those that die wash ashore where they can be examined. Given this evidence of fishing interactions around the main Hawaiian Islands, it is likely that they are also occasionally impacted by retaliation from some fishermen.

Long-term resightings and habitat use information from surveys and tagging data all indicate that there is a small island-associated population of pygmy killer whales around the main Hawaiian Islands, but only one population is recognized for all of Hawaiian waters. The smaller resident population, given its limited movements, small size, and overlap with a lot of human activities, is of greater conservation concern than the open-ocean population, but the lack of recognition of its existence means it is not taken into account in environmental planning—for example, recognizing or mitigating potential risks from high-intensity naval sonars. The problem is that conclusively proving there is a distinct resident population is going to be more difficult than for some of the common species, owing to the difficulty of tagging individuals or collecting genetic samples from different populations for comparison. So such recognition may not be coming soon.

MELON-HEADED WHALES *(Peponocephala electra)*

Charles Wilkes, commander of the U.S. Exploring Expedition, wrote the following entry in his report during the expedition's visit to Hawai'i Island in 1841:

> On the 12th of February, I witnessed an interesting sight,—the chase of blackfish, of which a school was seen in the afternoon in the bay. Upon this, the natives who were fishing, and those on shore, put off in their canoes to get to seaward of them: when this was effected, they began making a great noise, to drive the fish in; and finally succeeded in forcing many of them into shoal water, from whence they were dragged on the beach, when about twenty of large size were taken. . . . They afforded a fine feast to all the inhabitants of the bay, besides yielding plenty of oil. . . . The moment a school of porpoises is discovered, it is their usual practice to drive them in. (Wilkes 1845)

Wilkes was describing a group of small whales being herded into Hilo Bay. One animal was measured and a specimen was collected, and this would later become one of only four type specimens of the melon-headed whale. Whether such hunts had occurred prior to Europeans arriving in the islands is lost to history.

On July 3, 2004, a group of between 150 and 200 melon-headed whales was observed moving into the shallow waters of Hanalei Bay, on the north side of Kaua'i. This was an unusual sight, for melon-headed whales normally keep to deeper waters in Hawai'i. The whales moved out of the bay and then came back in, and they were seen in shallow waters milling and spyhopping in what would later be termed "pre-stranding" behavior. The reaction of local residents was quite different than in 1841: the next morning, community members, volunteers, the local stranding network, and others helped herd the whales out of the bay, where they had spent over twenty-eight hours. Only one melon-headed whale, a calf, was found dead the next day. What likely drove the whales into Hanalei Bay was a different type of "great noise." Prior to and after the whales were in the bay, naval vessels were offshore of the island to the north using MFA sonar as part of the RIMPAC military exercise. A review of the event concluded that naval sonar was "a plausible, if not likely, contributing factor" in the near mass stranding. By the time of the Hanalei Bay event, we had encountered melon-headed whales on

seven occasions around the islands, off Kaua'i, O'ahu, and Hawai'i Island, and we had taken over a thousand photos. After the stranding, to learn more about the species in Hawaiian waters and help understand the circumstances of the Hanalei Bay event, we established a photo-identification catalog of melon-headed whales, later to become the subject of a master's thesis by Jessica Aschettino of Hawai'i Pacific University.

Identifying Features and Similar Species

Melon-headed whales are one of the two smaller blackfish, born at about 1 to 1.15 m (3′3″ to 3′9″) long, and reaching maximum lengths of about 2.78 m (~9′1″) and a maximum weight of about 275 kg (~606 lbs). The largest animal measured from Hawai'i was 2.64 m (8′8″) long, one of the animals killed in Hilo Bay in 1841. They are generally dark gray to black in coloration. They have a darker dorsal cape that starts narrowly at the front of the head and extends backward as a narrow strip, then widens just in front of the dorsal fin and dips at a steep angle well below the fin. The boundary between the dorsal cape and the lateral pigmentation is diffuse and difficult to see except in good lighting conditions. They also have a dark mask on the side of the face, extending from around the eye to the front of the melon. When viewed from above they have a slightly pointed head, and from the side the head becomes more rounded in older individuals. Older animals also have lighter-colored or white lips. On the ventral surface, they have a lighter-colored blaze between the flippers. Scarring on melon-headed whales, including cookie-cutter scars, appears to repigment to the background color quite quickly. Adult males are slightly larger than adult females and have a distinct ventral keel that can be used to discriminate the sexes.

Melon-headed whales are most frequently confused with pygmy killer whales, but with good photos or in good lighting conditions it is relatively easy to distinguish the two species. The steep angle and diffuse boundary of the cape as

‹‹‹

A leaping melon-headed whale from the Kohala resident population, October 19, 2011. Photo by Jessica M. Aschettino.

[Top] A group of four melon-headed whales from the Kohala resident population, June 4, 2015. [Bottom] A melon-headed whale from the Hawaiian Islands population, showing the diffuse cape that extends steeply below the dorsal fin, as well as the facial mask and pointed flippers characteristic of this species. This individual, HIPe0789 in our catalog, was the second melon-headed whale we satellite tagged, shortly after this photo was taken off Kona, April 19, 2008. Over the nine days he was tracked, he moved 560 km to the northwest, with the last location west of Ka'ula Island. Photos by author.

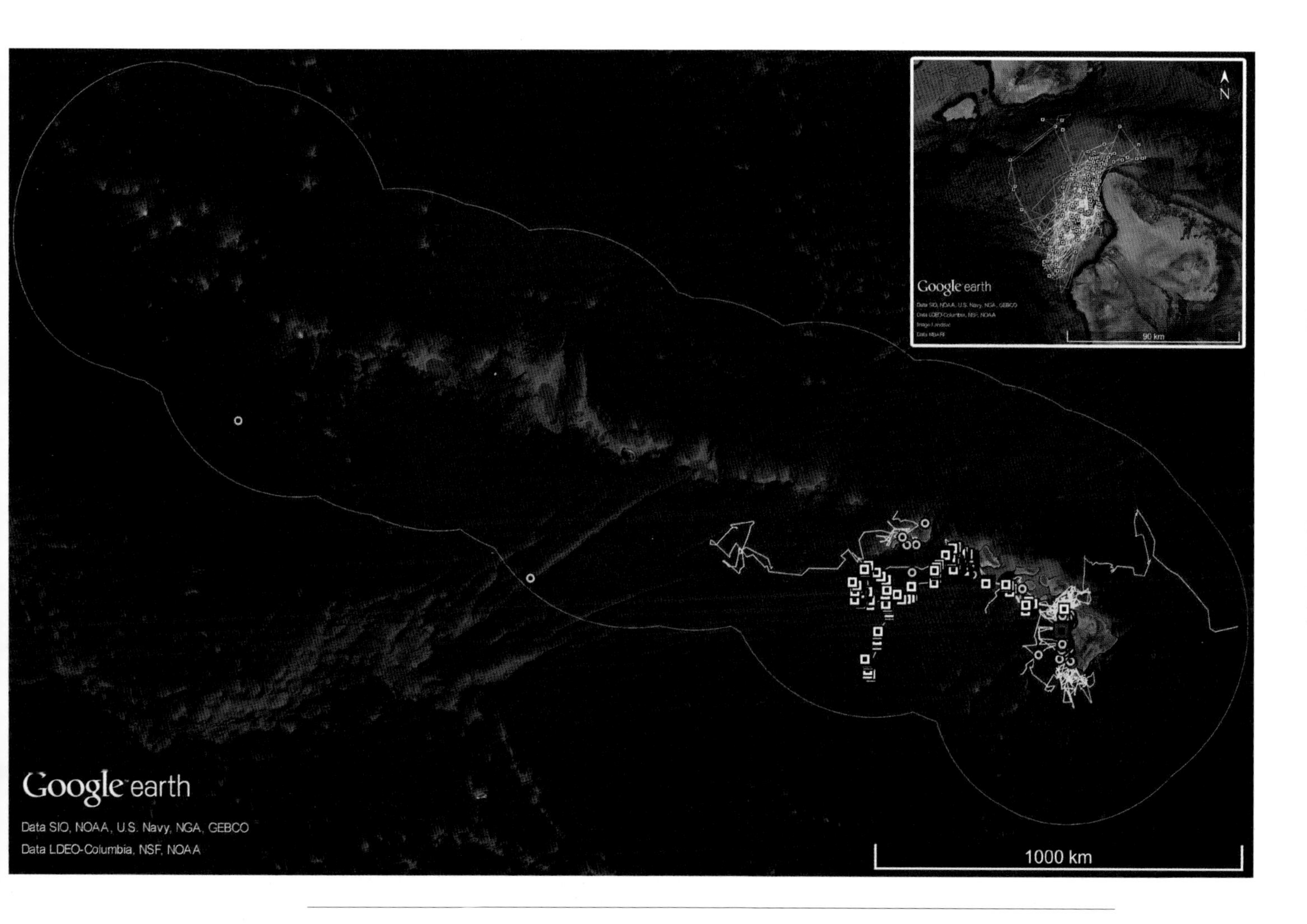

Melon-headed whale sighting records (yellow circles) and tracks from twenty-three satellite tagged individuals (lines). Locations from one individual from the Hawaiian Islands population tagged off Kona in April 2008 and tracked over a twenty-three-day period are indicated with white squares, with the initial tagging location shown in red. The inset shows tracks (lines) of eight individuals from the Kohala resident population tagged between 2008 and 2015. Locations of one individual from the Kohala resident population over a thirteen-day period in June 2015 are indicated by white squares.

it dips down is one of the easiest ways to tell melon-headed whales apart from pygmy killer whales, which have a much more clearly demarcated cape that dips down at a relatively shallow angle. If just the head is seen, the darker facial mask rather than the cap on the crown of the head is also a diagnostic feature. The flippers of melon-headed whales are pointed, whereas pygmy killer whales have rounded flippers. Melon-headed whale groups are usually much larger than those of pygmy killer whales, often hundreds of individuals, but small groups are also seen occasionally.

Habitat Use, Movements, and Abundance

Melon-headed whales are found in tropical oceanic waters worldwide and are typically found near shore only around oceanic islands. There are few tropical oceanic islands with as easy access to deep protected waters as there are in Hawai'i, and thus there are a lot more opportunities to learn about melon-headed whales here than elsewhere. Although we don't see them very often—just sixty sightings over a fifteen-year period—we've tried to make the most of these encounters. We have taken over 99,000 photos, collected 173 biopsy samples, and satellite tagged 23 individuals. From this work, more is known about melon-headed whales in Hawai'i than anywhere else in the world. Two populations have been identified here: a resident population off the Kohala area of Hawai'i Island (called the Kohala residents) and a Hawaiian Islands population. Individuals from the Hawaiian Islands population move among the islands, venture offshore, and often return back to the islands.

The Kohala residents have a very limited range, probably the smallest range of any recognized population of whales or dolphins in Hawaiian waters. We've satellite tagged eight different individuals from this population in four different years, and they remained in an area that ranges from just east and north of 'Upolu Point, the northernmost point of Hawai'i Island, to off of Keāhole Point in the middle of the Kona Coast. There has been one sighting of the Kohala residents as far south as the town of Kailua-Kona. All of the sightings and the locations from the satellite tags indicate that individuals from this population spend the majority of their time in waters between 300 and 1,000 m deep, only occasionally moving into the deeper waters of the channel between Maui and Hawai'i Island.

Individuals from the Hawaiian Islands population, by contrast, spend almost all their time in depths greater than 1,000 m, regularly move among islands and into offshore waters, and at least occasionally move into international waters. The ranges of the two populations do overlap off Hawai'i Island, but we've never seen individuals from both populations together in the same group, and preliminary genetic evidence suggests that they rarely, if ever, interbreed. Offshore movements can best be described as meandering; they tend not to go in anything resembling a straight line for very long. Analyses of the tag data in relation to the eddies that spin off Hawai'i Island show that sometimes the Hawaiian Islands individuals are paying attention to these eddies, foraging along the edges of upwelling (cold-core or cyclonic) eddies or in the center of downwelling (warm-core or anticylonic) eddies.

The Kohala resident population is thought to number between four and five hundred individuals, estimated from analyses of photo-identified individuals, while the Hawaiian Islands population is more than ten times larger, recently estimated at more than eight thousand individuals from the 2010 NMFS survey.

Predators and Prey

We've photographed a number of different melon-headed whales that had recently been attacked by large sharks, with major bite wounds on the dorsal fin and the back around the fin. As with pygmy killer whales and dwarf sperm whales, the locations of these bites suggest that melon-headed whales may roll over as a defensive mechanism if they detect a shark about to bite. From the frequency of wounds, either they are attacked more frequently than some other species of small delphinids or they just may have a better chance of surviving such attacks, although only one of these individuals has been resighted so far, with the wounds well healed. Melon-headed whales are probably attacked by killer whales in Hawaiian waters rarely, given how infrequent sightings of killer whales are, but the first time we saw killer whales in Hawai'i was near a group of melon-headed whales, with the melon-heads scattering in all directions, so they definitely viewed the killer whales as a threat.

Melon-headed whales have twenty to twenty-five pairs of fairly small teeth, which reflect in part their diet—they tend to feed on relatively small fish and squid. During the day we normally see melon-headed whales slowly traveling or resting at the surface in large groups—we've never seen any obvious foraging behavior. Three of the satellite tags we've deployed on melon-headed whales also transmitted information on dive depths and durations, providing information on day and nighttime behavior. During the day they rarely dove below 50 m, and the deepest dive was to 180 m, while at night such dives were common: one individual was documented diving to 472 m, staying down almost 12 minutes. So it appears that melon-headed whales in Hawaiian waters do most of their foraging at night, likely on prey that are associated with the deep-scattering layer that rises toward the surface at night.

Life History and Behavior

Female melon-headed whales are sexually mature at about seven years of age and give birth to their first calf at eight or nine years of age. Gestation lasts about a year, and they have one calf every three to four years. We've documented newborn individuals in eleven months of the year, all except December, suggesting that calving occurs year-round, but there is a peak in sightings of newborn animals in Hawaiian waters between March and June. Males are not sexually mature until their midteens. The oldest individual that has been aged was thirty-six years old, but they probably live longer.

Melon-headed whales are found in the largest groups of any species of whale or dolphin in Hawaiian waters. Although we have seen lone individuals on two occasions (one of which was injured and probably dying), the largest group we've seen had an estimated 800 individuals, and the average group size is almost 250. For the Kohala resident population, four of our sightings have been of groups of over 400 individuals, close to the estimated population size. This suggests that, at times, the entire Kohala resident population may be together in one large group. While this is not unlike what is often documented for the fish-eating "southern resident" killer whales in Washington State, this is the only population of whales or dolphins in Hawaiian waters that is known to have all members of the population potentially together in one small area at one time.

Like spotted dolphins, we sometimes see segregation of melon-headed whales by sex and age, with subgroups within a larger group composed of a number of

A group of melon-headed whales from the Hawaiian Island population logging at the surface off Kaua'i, June 30, 2008. Group size during this encounter was estimated at 340 individuals, and 25 individuals are visible in this photo. Photo by author.

A melon-headed whale from the Kohala resident population leaping, October 24, 2011. Photo by Jessica M. Aschettino.

adult males together or a number of females and calves. Melon-headed whales are one of a handful of whale species that have been recorded mass stranding, with a number of strandings of several hundred animals documented in various locations around the world. They regularly associate and intermingle with other species of whales and dolphins. More than a third of our sightings of this species have had other species present, particularly rough-toothed dolphins, but also short-finned pilot whales, humpback whales, and bottlenose dolphins. We've seen Fraser's dolphins only four times in Hawai'i, and during two of the sightings they were associated with groups of melon-headed whales.

The hole through the dorsal fin of this melon-headed whale was likely caused by a bite from a cookie-cuttter shark. This individual, part of the Hawaiian Islands population, was photographed off Kona, July 23, 2014. Photo by author.

Conservation

Probably the greatest threat facing melon-headed whales in Hawai'i comes from the fact that the range of the Kohala resident population is so small, and at least on occasion all individuals in the population may be together in one group. Based on the Hanalei Bay event—and a mass stranding of melon-headed whales that occurred in Madagascar associated with the use of a high-intensity multibeam sonar that was being used for ocean floor mapping—it is clear that melon-headed whales are one of those species that is easily disturbed by very loud sounds. Given the frequency of naval training operations in Hawai'i, it is not hard to imagine a scenario where all or most of the Kohala resident population could be exposed to high-intensity sonar and either be forced out of its normal range into unfamiliar waters or even end up stranding while trying to get away from the sonar. Until and unless high-intensity sonars are prohibited in the area surrounding the range of this population, off the northwest coast of the island of Hawai'i, I think they will always be at risk of a catastrophic event potentially affecting the entire population.

Melon-headed whales may also be affected by other types of human activities. Individuals from both the Hawaiian Islands population and the Kohala resident population have dorsal fin injuries suggestive of line entanglements, similar to what has commonly been documented for the main Hawaiian Islands false killer whales. Whether they are sometimes taking bait off fishermen's lines is unknown, but that is one potential source of such injuries. There are also five melon-headed whales that we have photos of that appear to have bullet wounds in the dorsal fins. I suspect that melon-headed whales may have been occasionally targeted as a result of their resemblance to false killer whales, and fishermen have mistakenly shot at them, thinking they were taking their catch. It is ironic, given that melon-headed whales feed only on small squid and deepwater fish and do most of their feeding at night.

KILLER WHALES *(Orcinus orca)*

Killer whales are the largest member of the dolphin family, and they are perhaps the most distinctive and easily recognized cetacean in the world. Most boaters and members of the public today would recognize this species, yet despite this, other than reports by whalers in the 1800s, the first confirmed record from Hawaiʻi was a stranding at Ka Lae, Hawaiʻi Island, in 1950. In recent years there have been usually only one or two sightings in Hawaiian waters each year, reflecting both their use of offshore waters around the islands and their relative rarity in the area.

Killer whales are one of the best-studied species of whales or dolphins in the world, with documentation of behavior, ecology, and genetics, among other things. Although guidebooks will note that there is a single species of killer whale worldwide, all the evidence suggests there are at least two and probably more species that have just not yet been formally described. The two that have been reproductively isolated for the longest period, hundreds of thousands of years, are the well-known forms along the western coast of North America—the fish-eating so-called residents and the mammal-eating so-called transients—and these two behave as distinct species. Two of the "southern resident" killer whales, captured near Seattle in Washington State in 1968, were actually brought to Hawaiʻi by the U.S. Navy. Both of the whales were being trained for use in a deepwater object recovery program in 1971 when one of them, a male named Ishmael, decided to go AWOL off Oʻahu. Whether Ishmael survived and had the opportunity to contribute to the genetic makeup of killer whales in Hawaiian waters is unknown, but if he was able to mate, his impact was probably small, as killer whales in Hawaiian waters do not appear to be either of these types. Instead they differ morphologically as well as behaviorally, suggesting they are part of a distinct population found throughout the central tropical Pacific.

Identifying Features and Similar Species

Killer whales have a striking black and white pattern that is difficult to confuse with any other species, including a bright white oval patch above and behind the eye, a gray "saddle" patch behind and below the dorsal fin, a white throat,

A killer whale off Kona carrying a bigeye thresher shark, November 2, 2013. Photo by Deron S. Verbeck/iamaquatic.com.

Killer whales are strongly sexually dimorphic, both in body size (males are larger than females) and in the relative size of the dorsal fin, tail flukes, and pectoral flippers. Male (top); female (bottom). Illustrations by Uko Gorter.

partially white belly and flanks, and white on the underside of the tail. The saddle patch on killer whales in Hawaiian waters is less distinct than on individuals from the so-called resident or so-called transient populations, and it is only really obvious in good lighting conditions. Killer whales are sexually dimorphic in both body size and appendage size, with adult males larger than adult females, and the dorsal fin, pectoral flippers, and tail flukes of adult male killer whales are all substantially larger in proportion to the body size than for adult females. The dorsal fin in particular can be twice as tall in an adult male as in an adult female. The tail

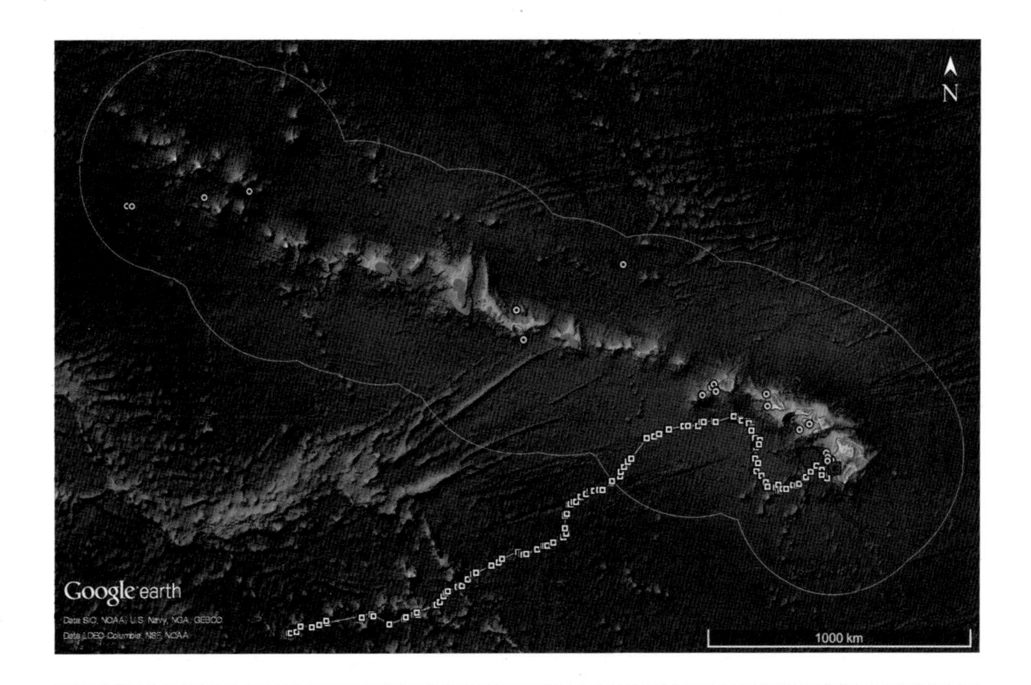

Killer whale sighting records (yellow circles) along with the track and locations (white squares) of an individual killer whale satellite tagged off Hawai'i Island on November 1, 2013. The tagging location is indicated with a red symbol. Three individuals were tagged in the group and remained together. Over the twenty-five-day tracking period, they moved almost 2,000 km to the southwest, more than halfway toward the Marshall Islands in the western Pacific.

flukes of older adult males also tend to curve downward at the tips. The sexual dimorphism, at least in terms of the dorsal fin size, appears to be less extreme for killer whales in Hawaiian waters; the dorsal fin of adult males does not appear to be as obviously tall and straight as it does for killer whales elsewhere. That said, when an adult male is present the taller dorsal fin is diagnostic—there is no other species in Hawaiian waters that has such a large dorsal fin in relation to body size.

While killer whales are the largest member of the dolphin family, with individuals in some populations reaching lengths of 8 m (~26′) or more, individuals in Hawaiian waters appear to be relatively small. One animal documented in an aerial survey off Ni'ihau in March 2000 was estimated using photogrammetry at 6.5 m (21´4″) long, but an adult male that stranded and died on Kaua'i in 2008 was measured at about 5.5 m (18´) long. One adult male that we closely approached in our work was estimated at only about 5 m (about 16´5″) long, appearing smaller than some of the larger adult male pilot whales we've encountered.

Habitat Use, Movements, and Abundance

Killer whales have been documented off the Northwestern Hawaiian Islands, in offshore waters, and around the main Hawaiian Islands, although most sightings around the main islands have been relatively far from shore. In our work, with over two thousand sightings of odontocetes around the islands through the end of 2015, we had encountered killer whales on only three occasions, once in the channel between Kaua'i and Ni'ihau (in July 2011) and twice off Hawai'i Island (in May 2003 and November 2013). Combining all the photos we've been able to obtain from other researchers and members of the public, we have only thirty-two individuals in our Hawai'i photo-identification catalog, and with the exception of one group seen two days in a row, none have ever been resighted. Our photographs have been compared with an eastern tropical Pacific killer whale catalog held at the Southwest Fisheries Science Center, but there are no matches with that catalog. In our November 2013 encounter, we were able to deploy satellite tags on three individuals and track the movements of the group over a twenty-five-day period. This was the first time killer whales in the tropics had been satellite tagged. By the end of the tracking period they had moved almost 2,000 km from where they were tagged, following a meandering path and not approaching the islands again, bypassing Johnston Atoll to the southwest, and spending most of their time in water depths ranging from about 4,500 to almost 6,000 m deep. By the time the tags stopped transmitting, the whales were halfway to the Marshall Islands, in the western Pacific. During the time of overlap between the three tagged whales, all stayed close to each other.

Combining the information from photo identification, sightings, and our one satellite tagged group, we can say a few things about killer whales in Hawaiian waters. Unlike the majority of the other delphinids in Hawaiian waters, killer whales do not appear to have a resident population around the main Hawaiian

A subadult killer whale off Kona, November 1, 2013. Although the ribs are showing, this individual is actually quite robust. The groove down the middle of the back behind the blowhole reflects the large muscle masses built up on either side of the midline. Like many killer whales in the tropics, this individual has a number of hitchhiking remoras attached. A LIMPET satellite tag is visible on the dorsal fin. Photo by author.

Islands; those that are seen around the islands are likely part of a widely ranging open-ocean population. Given the morphological differences (small body size, very indistinct saddle patch, reduced sexual dimorphism), this population is probably restricted to the central tropical Pacific, not extending to coastal areas to the north or east, where multiple studies of killer whales have documented animals that are quite different in appearance.

Killer whales are top predators, and top predators (think lions, tigers, and bears) are naturally rare. There are two abundance estimates available for killer whales in Hawaiian waters: about 350 individuals from the 2002 NMFS survey and about 150 individuals from the 2010 NMFS survey. The estimate from the 2010 survey is the lowest for any species of odontocete in Hawaiian waters.

Predators and Prey

Killer whales certainly might be at some risk of attack from large sharks such as white or tiger sharks, but the reverse is probably more likely the case. While killer whale populations in many coastal areas are known to specialize on certain prey types, such as fish (particularly Chinook and chum salmon) along the west coast of the United States and Canada, in Hawaiian waters they have been seen attacking a diversity of prey. Surprisingly, there has been only one documented attack on a humpback whale calf, and there does not seem to be a particular peak in sightings of killer whales during the humpback whale calving and breeding period, suggesting that they don't seek out the islands when large numbers of humpback calves are available. In our July 2011 sighting between Kaua'i and Ni'ihau, the whales were first seen chasing a rough-toothed dolphin, and in our May 2003 sighting the first indication of killer whales being present was a large group of melon-headed whales scattering when killer whales appeared in the middle of them. Combined with sightings of an attack on pantropical spotted dolphins, it is clear that killer whales in Hawai'i do at least occasionally attack other species of toothed whales. But killer whales in Hawai'i do not seem to focus just on marine mammals. One killer whale that stranded on Lāna'i in 2004 had a number of squid in the stomach, one group was seen feeding on an "octopus," and there have been several observations of them feeding on large sharks, including both a hammerhead and a thresher shark. So all the evidence suggests that killer

A killer whale carrying a bigeye thresher shark off Kona, November 2, 2013. Photo by Russel D. Andrews.

whales found around the Hawaiian Islands are generalists, not specialists, unlike the better-studied populations along the west coast of North America. This is not surprising, really, given the overall low productivity of the waters around the Hawaiian Islands. Specializing on a certain prey type makes sense only where there are enough of one type of prey to make it worthwhile; in areas with low productivity such as Hawai'i, being a generalist is a much better strategy. As generalists, they would probably attack a Hawaiian monk seal if they came across one, but such attacks are likely extremely rare given that killer whales in Hawai'i tend to be found offshore in deeper waters, while monk seals are usually closer to shore unless they are transiting between islands. This is obviously a good thing for Hawaiian monk seals, as they are critically endangered.

Life History and Behavior

Like the other blackfish, killer whales take a long time to mature—females are probably not sexually mature until ten to twelve years of age—and are slow to reproduce. Calving intervals for the so-called resident populations along the west coast of the United States and Canada tend to be about one calf every five years, although it is not known what the calving interval might be for killer whales in the less-productive central tropical Pacific. They are one of several species of whales whose females go through menopause and stop reproducing, usually by their late thirties or early forties, but continue to live for a long time afterward. Some females may live into their eighties or even nineties. Males, on the other hand, live much shorter lives—their maximum longevity is closer to fifty years.

Killer whales are well known for their strong and enduring social bonds, with individuals in some populations remaining together for life. Group sizes in Hawai'i are relatively small; lone individuals are occasionally seen, and the largest reported group was of ten individuals, with average group sizes of about five individuals. Most of the smaller groups, two to four individuals, likely represent a mother and her offspring. Like the so-called transient killer whales off the west coast of the United States and Canada, males may stay with their mothers for life, while females probably disperse and form their own groups when they are sexually mature. Most sightings of larger killer whale groups in Hawai'i probably represent temporary associations of individuals that come together for foraging.

On November 1, 2013, we encountered a group of four killer whales off Kona and satellite tagged three. Using information from the tag data, Russ Andrews relocated the group the next day, and there were seven individuals present; that day they were feeding on a large bigeye thresher shark. Cooperative hunting—the practice of hunting with familiar hunting partners when trying to attack dangerous or difficult-to-capture prey—is likely one of the main reasons killer whales form such stable groups.

With only three encounters of killer whales in Hawaiian waters, it is difficult to generalize how this species behaves toward boats. During our first encounter,

A killer whale off Kona, November 1, 2013. This individual can be identified as an adult male by the large size of the appendages and the downward curvature of the tail flukes. The saddle patch is indistinct, typical of killer whales in the central tropical Pacific. Photo by Deron S. Verbeck/iamaquatic.com.

off Hawai'i Island in 2003, the group appeared to avoid our vessel, although just before we approached, another boat had rapidly approached the group at close quarters, and perhaps we were considered guilty by association. During our second encounter, off Kaua'i in 2011, the group approached our boat as if curious but then went on a long dive and was not seen again. Our most recent encounter, off Hawai'i Island in November 2013, was the opposite of the first, with three of the four individuals showing considerable interest in our boat and no avoidance. Several photographers on another boat got in the water with the whales on that occasion, and the whales were largely indifferent to the people in the water.

Conservation

Top predators are typically more at risk from impacts of human activities than other species for a couple of reasons. Top predators are naturally rare, and with smaller numbers of individuals they are more subject to the impacts of inbreeding. If populations decline due to some human activity or environmental factor, there is less potential for recovery. In the case of killer whales and other top predators such as false killer whales, as they feed at the top of the food web they accumulate higher levels of persistent organic pollutants such as pesticides (for example, DDT) and industrial chemicals such as PCBs and flame retardants. High levels of contaminants, particularly PCBs, are known to cause suppression of the immune system and other health issues. The male killer whale that stranded on Kaua'i in 2008 had the highest levels of pesticides and PCBs of any of the forty-two individuals of sixteen different species from Hawai'i analyzed in a recent study. This individual had a variety of adrenal and kidney lesions, and the authors of that study noted that the high levels of contaminants may have contributed to the animal's poor health.

COMMON BOTTLENOSE DOLPHINS *(Tursiops truncatus)*

In 1957 and 1958, biologist Dale Rice documented bottlenose dolphins in the Northwestern Hawaiian Islands and posed a question that was only recently answered: "Are the populations local and isolated, or migratory and intermingling? The infrequency of observations far at sea would favor the former hypothesis." Bottlenose dolphins are the main reason that a yearlong job for me on Maui in 1999 turned into a seventeen-year (and counting) obsession with understanding the resident whales and dolphins in Hawaiian waters. Bottlenose dolphins are commonly seen around Maui and Lāna'i, are easily approached and photographed, and most individuals are quite distinctive, making them easy to recognize. So, when I first started working on Maui in 1999, I immediately began compiling a photo-identification catalog of this species, with the help of Annie Gorgone, a friend and colleague. Prior to that, virtually nothing was known about bottlenose dolphins in Hawaiian waters—no one had tried to answer Dale Rice's question. Our photo-identification work off Maui and Lāna'i in early 2000—funded ironically and indirectly by a small grant to study humpback whale diving behavior—led to a contract to come back to the islands the following winter to obtain more photos and estimate the abundance of bottlenose dolphins. Our estimate of bottlenose dolphins using the Maui Nui Basin was only 134 individuals, a surprisingly small number. Once we had an estimate of abundance, I was able to use that to get another contract to come back in April and May of 2002, this time to learn whether any of the bottlenose dolphins from Maui Nui were moving to O'ahu or to Hawai'i Island. That project led to a long-term collaboration with Dan McSweeney in Kona, working with many different species, and it resulted in another grant the next year to revisit all those islands and expand the work to Kaua'i and Ni'ihau. Today we have a catalog of over six hundred distinctive bottlenose dolphins from among the islands.

While bottlenose dolphins are found almost worldwide, there are at least two species and a number of forms that are recognized, including several from the North Pacific. One collected in 1906 in the eastern tropical Pacific was described in 1911 as *Tursiops nuuanu,* named after the sailing vessel *Nuuanu* from which it was collected (Nu'uanu, on O'ahu, is the site of a famous battle led by Kamehameha I in 1795). While *Tursiops nuuanu* was later lumped into the

A group of bottlenose dolphins off Kona, February 9, 2014. Photo by Deron S. Verbeck/iamaquatic.com.

common bottlenose dolphin (which I refer to simply as the bottlenose dolphin), it is certainly possible that if more species of bottlenose dolphins are recognized, one may again get a Hawaiian name. While there is only one species of bottlenose dolphin recognized from Hawaiian waters, four individuals we've biopsied off Kaua'i have evidence of hybridization with Indo-Pacific bottlenose dolphins (*Tursiops aduncus*), a species thought to be restricted to the far western Pacific, with the closest known population from the Bonin (Ogasawara) Islands, south of Japan. Whether individuals have traveled from the Bonin Islands or there is some small unrecognized population of Indo-Pacific bottlenose dolphins hiding somewhere in Hawaiian waters (such as near Midway or Kure Atolls) is unknown and will have to await future genetic analyses of samples collected.

Identifying Features and Similar Species

Describing a "typical" bottlenose dolphin in Hawai'i is a bit difficult, given that there are two different forms—a large offshore type and a smaller nearshore island-associated type—and at least four different populations of the smaller island-associated type, with some variability in scarring patterns among them. Bottlenose dolphins are a midsized dolphin, born at about 1 to 1.3 m (3′3″ to 4′3″) in length, and reaching maximum lengths of about 3.8 m (12′6″). Offshore animals are both larger and more robust than individuals from the island-associated populations, and have a particularly robust beak. The largest animal measured in Hawaiian waters was a 2.91 m (9′6″) individual that stranded on O'ahu in 2004, but adult pelagic animals probably reach 3–3.5 m (9′10″–11′6″) in Hawai'i. They get their name from their short snout or rostrum, which is quite robust and clearly demarcated from the melon. They are generally gray in color with a slightly darker gray cape and a white belly, and older animals are often covered with extensive, small, light-colored linear scars from tooth rakes by other dolphins. Scars from cookie-cutter shark bites appear a lighter gray in color, rather than the white scars seen on species such as beaked whales and rough-toothed dolphins. Adult bottlenose dolphins in the Maui Nui Basin have few scars from cookie-cutter sharks, as they appear to spend most of their time in very shallow water. Adults off other islands tend to have more scars, as they do occasionally use deeper waters where cookie-cutter sharks are found.

Bottlenose dolphins off Kaua'i, February 2, 2013. Photo by Brenda K. Rone.

Individuals from the offshore population have extensive scarring from cookie-cutter shark bites.

If there is a population of Indo-Pacific bottlenose dolphins in Hawaiian waters, this is clearly the species with which they would most likely be confused, and the two species would be very difficult to tell apart. Indo-Pacific bottlenose dolphins tend to have a longer beak and are more slender, and they have a relatively larger dorsal fin. Most adult Indo-Pacific bottlenose dolphins also have black spots or flecks on the belly. The distinct but short rostrum of bottlenose dolphins should make them relatively easy to distinguish from other nearshore dolphins such as spinner or spotted dolphins, or when they overlap with rough-toothed dolphins when they come close into shallower waters, such as off Kaua'i.

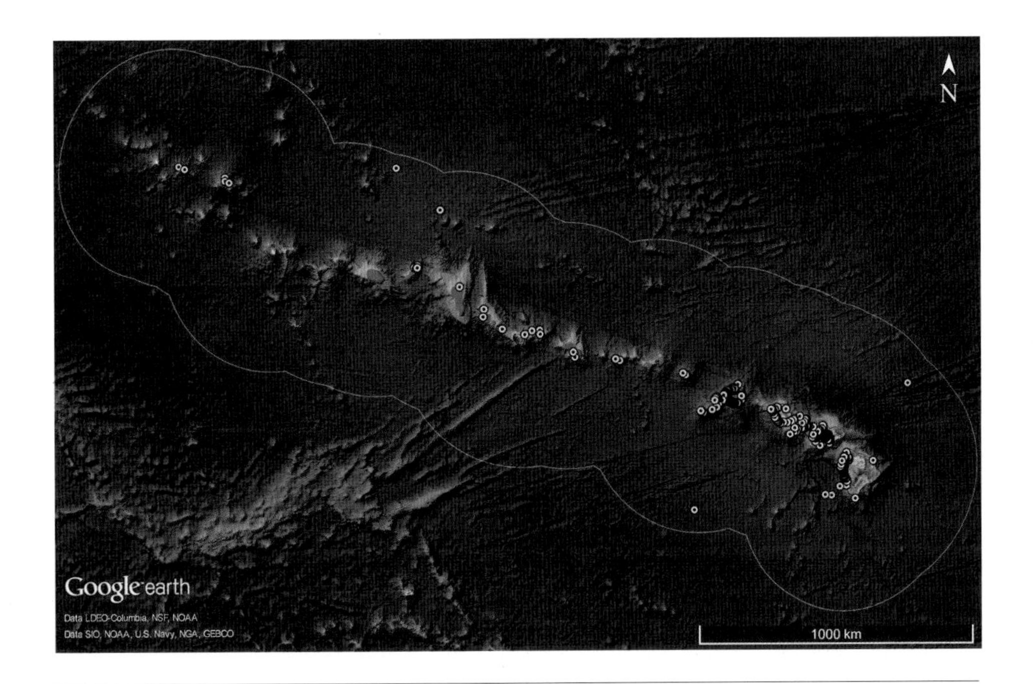

Bottlenose dolphin sighting records in Hawaiian waters. Almost all the sighting records are around the islands, reflecting the relatively low density of this species in pelagic waters in Hawai'i.

Habitat Use, Movements, and Abundance

Bottlenose dolphins are found both in the open ocean and in nearshore areas throughout tropical and temperate waters worldwide. In Hawaiian waters, bottlenose dolphins are seen throughout the Northwestern Hawaiian Islands and around the main Hawaiian Islands, but there are only a few sightings in offshore waters. Around the main Hawaiian Islands, they are found almost exclusively in depths less than 1,000 m. As of January 2016 we've had 292 sightings of bottlenose dolphins over the years, and all but three of them were in less than 1,000 m depth. Of the three in depths greater than 1,000 m, two were in slightly deeper water off the steep Kona Coast and were individuals that have been photographed before or since in shallow nearshore waters, part of the resident population. The only truly offshore group we've seen, a dozen individuals in almost 3,800 m depth and some 30 km offshore, were massive animals, estimated at 3.3–3.7 m (11–12′) in length. Despite our spending hundreds of hours searching in offshore waters, that is the only pelagic group of bottlenose dolphins we've seen in Hawai'i.

In less than 1,000 m depth, bottlenose dolphins are the most frequently encountered odontocete in our work, although admittedly if we surveyed only within 100 m of shore we'd probably find spinner dolphins more often, given their tendency to rest in very shallow water during the day. In fact, spinner dolphins are the only species that shows a similar strong preference for shallow waters in Hawai'i. Our early trips to other islands in 2002 and 2003 found no evidence that individuals moved among the islands, and thousands of photos later, combined with genetic analysis of 146 biopsy samples collected from among the islands, it is clear that bottlenose dolphins in Hawai'i are homebodies. Those found around Hawai'i Island live there year-round and don't appear to move to Maui Nui. The same goes for those found off Maui, Moloka'i, and Lāna'i—dolphins move back and forth among those islands but not apparently elsewhere. We have documented two individuals that moved from O'ahu to Moloka'i, but they were not seen with any of the Maui Nui Basin residents, so they may not represent dispersal as much as individuals on a short walkabout.

Genetic analysis of skin biopsy samples we've collected show that there are four distinct island-associated populations in the main Hawaiian Islands: one off Kaua'i and Ni'ihau, one off O'ahu, one in the Maui Nui Basin, and one off Hawai'i Island. We are beginning to get information on short-term movements from satellite tags deployed on bottlenose dolphins as well; individuals we've tagged off Hawai'i Island and Lāna'i have stayed in those areas, with some movements to both the eastern and western sides of the islands. We've tagged thirteen bottlenose dolphins off Kaua'i, and twelve of them stayed around the coast, with many regularly circumnavigating the island. The thirteenth dolphin stayed off Kaua'i for nine days and then went on a walkabout. That individual took a day to cross the deepwater channel to O'ahu, and then it spent the next week south of O'ahu before the tag stopped transmitting. It is not known whether any of its Kaua'i companions went with it or what it was doing off O'ahu, but hopefully we'll get a photo of this individual at some point in the future and be able to determine whether we documented an animal moving between populations or just one on a rumspringa.

As you'd expect for animals that are found only in shallow water around steep-sided oceanic islands where shallow water is in short supply, the abundance estimates for each of the resident populations of bottlenose dolphins are

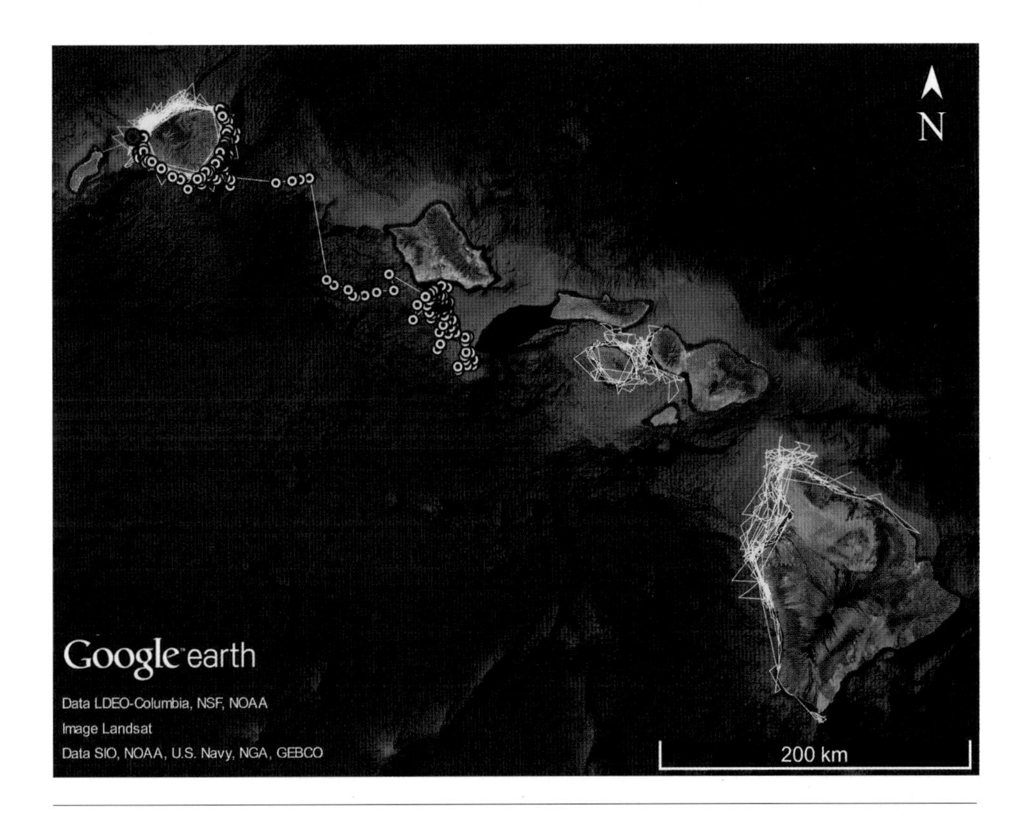

Data from seventeen satellite tagged bottlenose dolphins around the main Hawaiian Islands show that they rarely move offshore or across deepwater channels among islands. Locations are shown with yellow circles for one individual tagged off Kaua'i in October 2014 and tracked over fifteen days, with the tagging location indicated by a red symbol. This is the only tagged individual we've had move across one of the deepwater channels, but whether it remained off O'ahu is unknown.

relatively small, on the order of 150 to 250 individuals for each of Kaua'i/Ni'ihau, Maui Nui, and Hawai'i Island, while O'ahu seems to have a larger population, estimated at between 700 and 800 individuals. There is an estimate for the pelagic population based on the 2010 NMFS survey of over 20,000 individuals, but it is likely this estimate is too high, given that most of the sightings during that survey were of animals around the Northwestern Hawaiian Islands, and the estimate may have been capturing both an island-associated population there and a truly offshore population.

Predators and Prey

Bottlenose dolphins are at least occasionally attacked by large sharks in Hawaiian waters, based on half a dozen living animals that have healed shark-bite wounds. Attacks by killer whales in Hawaiian waters are probably even rarer than for most other species, as killer whales tend to stay in deep waters in Hawai'i, while bottlenose dolphins are in shallower waters.

We have seen bottlenose dolphins feeding during the day on a variety of nearshore and reef fish, as well as occasionally on small schooling fish. From diving data obtained from depth-transmitting satellite tags, they do dive deep much more often at night than they do during the day, so they seem to be feeding both during the day and at night. Some of the feeding dives are quite deep—the deepest we've documented a bottlenose dolphin diving to in Hawai'i was 752 m—so they likely bring only a small subset of their prey up to the surface.

Life History and Behavior

Bottlenose dolphin females become sexually mature between about five and thirteen years of age, while males mature later, between nine and fourteen years. Females can live more than fifty-seven years, while males may reach forty-eight. Gestation lasts about a year, and females have one calf every three to six years. In Hawai'i bottlenose dolphins appear to give birth year-round—we've seen newborn animals in all seasons.

Hawai'i bottlenose dolphins have been documented in groups of up to sixty individuals, but the average group size is only six. These animals are thought to live in a fission-fusion society, with groups mixing and matching sometimes based on age or sex. Stable associations are sometimes documented, particularly pairs of adult males. From analyses of association patterns in Hawai'i, it appears there may be differences between islands, possibly with larger, more cohesive groups of individuals off Hawai'i Island, although this needs further study.

Although most dives of bottlenose dolphins are in the three-to-six-minute range, we have documented five of eight individuals with depth-transmitting

A bottlenose dolphin with a damaged rostrum off Kona, July 5, 2008. This individual, HITt0150 in our catalog, was first documented off the island in 2002 and was seen as recently as 2012, so the damaged rostrum does not appear to be impairing feeding. Purple-stalked barnacles are protruding from the mouth. Photo by author.

tags diving more than ten minutes at a time, with the longest dive recorded of twelve minutes. These dolphins often show interest in boats, regularly bowriding and commonly approaching divers. In January 2013 during a night dive off Kona, one bottlenose dolphin that had fishing line entangled around its flipper approached a diver and allowed him to cut the line free—obviously it knew the diver could help.[2]

Bottlenose dolphins can be quite acrobatic, leaping far above the water, typically when engaged in social activities or when trying to dislodge remoras (suckerfish). They often associate with other species of whales or dolphins, in particular humpback whales during the winter. They seem to go out of their way to harass adult humpbacks or play with juvenile humpbacks when the opportunities arise. We've seen them interact with the main Hawaiian Islands false killer whales, occasionally apparently socializing with them and at least once trying to feed on a fish that a false killer whale had caught. Unlike cases where false killer whales will share their prey with humans, in this case they seemed to keep it away from the bottlenose dolphin. We've also seen them associated with pantropical spotted dolphins and, on one occasion each, with pilot whales, rough-toothed dolphins, and melon-headed whales. One bottlenose dolphin appeared to be harassing the melon-headed whales.

Conservation

Thirty-seven bottlenose dolphins were captured in Hawaiian waters, most off O'ahu, and taken into captivity in the 1960s and 1970s. Five bottlenose dolphins originally captured in the Atlantic were released into Hawaiian waters, including three U.S. Navy dolphins that were lost during training and two captive dolphins from the University of Hawai'i's Kewalo Basin Dolphin Laboratory that were stolen/liberated by ex-employees. Whether any of these survived and contributed to the genetic makeup of bottlenose dolphins in Hawai'i today is unknown.

The four resident populations of bottlenose dolphins in Hawai'i were recognized as distinct stocks by NMFS in 2010, which allows for assessing any impacts on a stock-by-stock basis rather than for all bottlenose dolphins in all of

2. A video of this interaction is available on YouTube at https://www.youtube.com/watch?v=XVMmPjwxESg.

A bottlenose dolphin investigating the Kona Kampachi fish farm off Kona, June 5, 2014.
Photo by Deron S. Verbeck/iamaquatic.com.

Hawaiian waters. Because of their nearshore habits and their roaming patterns along the coastlines of the islands, they are the species of dolphin in Hawai'i most likely to get entangled in gill nets, which are set close to shore in some areas in the main islands. At least two bottlenose dolphins have drowned in gill nets off Maui, although gillnetting is now banned around that island, as well as parts of O'ahu. Bottlenose dolphins do take bait and catch from fishermen on occasion, although less so than rough-toothed dolphins in Hawai'i, and animals sometimes

A bottlenose dolphin hooked on a shallow-set longline in international waters north of the Hawaiian Islands, February 29, 2012. An NMFS observer on board the vessel noted that this individual broke free from the line with the hook and part of the leader attached. Photo taken by NMFS Pacific Islands Region fishery observer David Tautofi.

get hooked or are shot at. How often such hookings or shootings occur is unknown, as are the outcomes, but given the small population sizes around most of the main Hawaiian Islands, it wouldn't have to happen too often to affect a population. When an aquaculture operation was opened off the Kona Coast, with net pens full of fish anchored just offshore, bottlenose dolphins began investigating the pens. Divers would regularly enter the pens and fish would sometimes escape, easy pickings for the dolphins. Some of the staff were also seen giving fish handouts to the dolphins. As well as being illegal, nothing good can come from this type of interaction; all it does is teach the dolphins that humans may be a source of food, leading to increased interactions with boats and divers. In other parts of the United States and elsewhere in the world where these types of interactions have happened, they don't turn out well for the dolphins.

Although bottlenose dolphins have been hooked and injured or killed in the offshore longline fisheries, such interactions are relatively rare compared to other species, such as false killer whales or Risso's dolphins.

ROUGH-TOOTHED DOLPHINS *(Steno bredanensis)*

It is rare for anyone other than scientists and aficionados to refer to a species of whale or dolphin by its scientific name, but off the island of Hawai'i rough-toothed dolphins are commonly called "Steno" by fishermen and the boating community. I think the use of the genus as a common name in Hawai'i arose in the 1960s and 1970s when biologist Ken Norris started his groundbreaking work with spinner dolphins. Asked by fishermen what type of dolphins were taking their bait or catch, I suspect Ken just referred to them as Steno, and the name stuck. Rough-toothed dolphins are typically considered an open-ocean species, but Hawaiian waters, particularly off Kaua'i, may be one of the best areas to see rough-toothed dolphins in the world.

The first record of a rough-toothed dolphin from Hawaiian waters was a specimen collected off Honolulu in 1846 as part of the Danish Galathea Expedition. In May 1964, two rough-toothed dolphins were captured off Wai'anae and taken to Sea Life Park. These were the first rough-toothed dolphins ever taken into captivity. Karen Pryor, the head trainer at Sea Life Park, notes in her 1975 book *Lads before the Wind* that the captive rough-toothed dolphins "have a tremendously long attention span and love a puzzle. They will sometimes go on working when they are too full to swallow another fish, just because the task is interesting". In February 2015 we witnessed puzzle solving by a group of rough-toothed dolphins off the east side of Kaua'i. The dolphins were swimming around some submerged netting that had attracted a school of small fish. From underwater video, we could see individual dolphins swimming close by the netting and snagging parts of it on their dorsal fins, dragging it away from the main mass of netting and then letting it loose. They appeared to be trying to separate the ball of netting to get at the school of fish hiding in the middle. Whether they were hungry or not is impossible to say, but as in Karen's description, the dolphins appeared to be methodically figuring out how to get the netting apart to get access to the fish.

Identifying Features and Similar Species

Rough-toothed dolphins are one of the smaller delphinids in Hawaiian waters. They are born at about a meter (3′3″) in length and may reach lengths up to

A group of rough-toothed dolphins off Kona, July 3, 2014. This species is commonly referred to as Steno by fishermen in Hawai'i. Photo by Deron S. Verbeck/iamaquatic.com.

Rough-toothed dolphins can be identified by their gently sloping forehead and unusually large flippers, and individuals are identified by the markings on the dorsal fin and pigmentation patterns. This photo was taken off Kona on October 29, 2009, but one of the individuals was first identified in 2004 and the other in 2008. Photo by author.

about 2.8 m (9′2″) and weights up to about 155 kg (~340 lbs). In his 1974 book *The Porpoise Watcher,* Ken Norris writes that he was "struck by the resemblance of these Stenos to extinct ichthyosaurs, seagoing reptiles from the age of dinosaurs. Its long snout lined with stout, pointed teeth, the big brown protruding eyes, and the reptilian head contours are all part of this impression." They are probably most easily distinguished by their sloping forehead, with no clear crease or demarcation of the beak. Their coloration is the typical dolphin countershading, with darker gray dorsal surfaces and lighter-gray sides. When they are born their lower

A piebald rough-toothed dolphin off Kaua'i, November 14, 2005. Photo by Annie B. Douglas.

jaw is gray, but as they age it becomes progressively whiter. Superimposed on the lighter- and darker-gray background pigmentation, rough-toothed dolphins have a blotchy appearance, and these patches (lighter on the dark dorsal surface and darker on the light lateral surface) appear to be long-term, possibly permanent pigmentation features that can be extremely useful for matching individuals over time, similar to the pigmentation of blue whales or humpback whale flukes. Like beaked whales, rough-toothed dolphins tend to become progressively more spotted as they age, with white oval spots the result of healed cookie-cutter shark

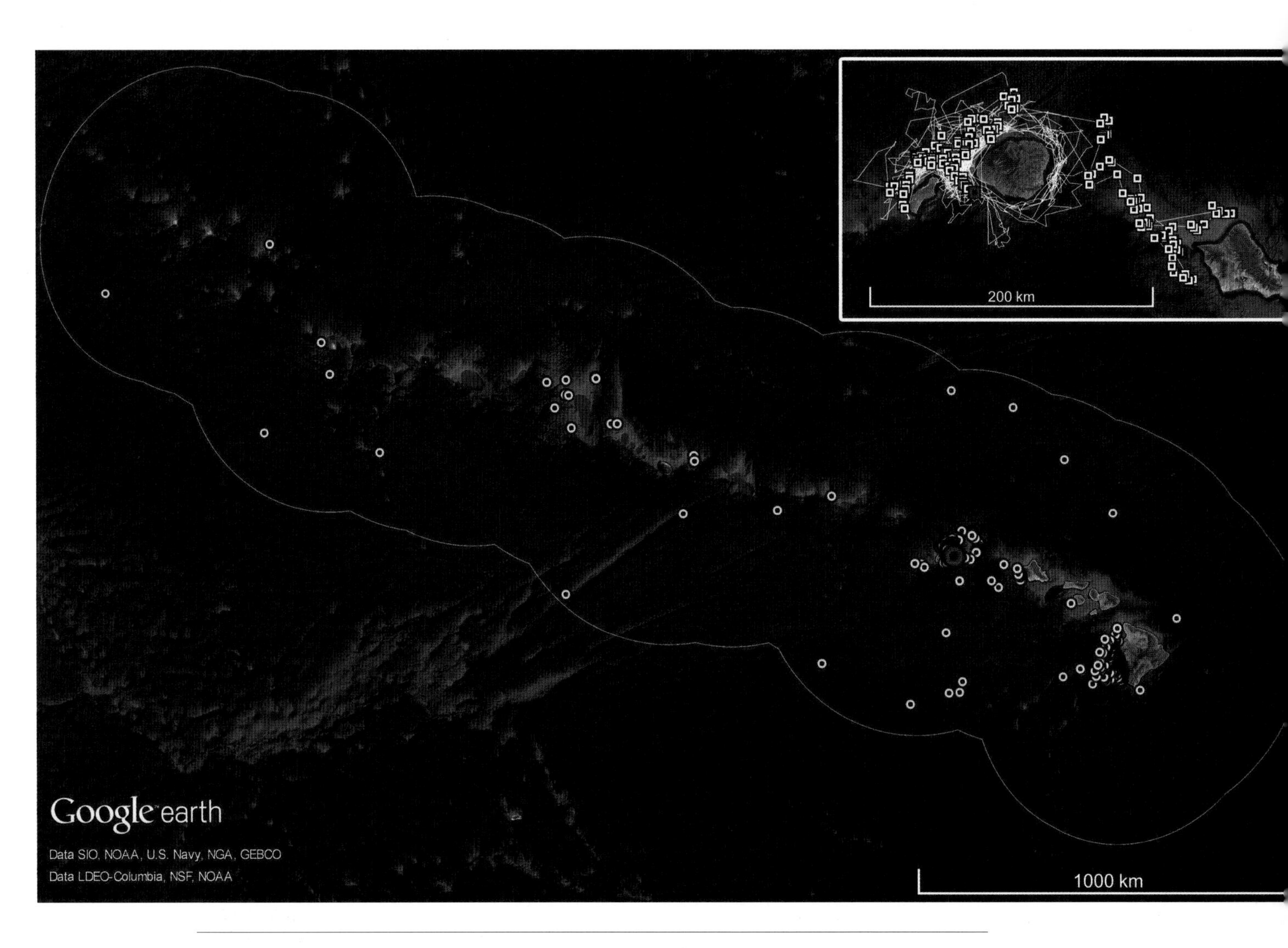

Sightings of rough-toothed dolphins in Hawaiian waters. The inset map shows tracks (white lines) from fourteen individuals satellite tagged off Kaua'i from 2012 through mid-2015. Locations are shown (white squares) for one individual tagged in February 2014 and tracked over a thirteen-day period, with the tagging location indicated by a red symbol.

bites, and the bellies in particular are often so heavily scarred they appear largely white or pink, albeit with very irregular patterns and edges. We've also seen five different piebald individuals, with varying amounts of blotchy white pigmentation patterns on the dorsal fin and body, making them very distinctive. While their common name comes from ridges on the teeth, these ridges are so small as to be largely indistinguishable. Adult males are on average a bit larger than adult females, and females have a longer rostrum than males, although neither of these can be easily distinguished in the field. Adult males do have a distinct ventral keel that is visible underwater or if an animal leaps clear of the water.

The species that appears most similar to rough-toothed dolphins and the most likely to be confused with them are bottlenose dolphins, but they are usually seen in much shallower water in Hawai'i than rough-toothed dolphins. If the head is seen, the clear demarcation of the beak (for bottlenose dolphins) or lack thereof (for rough-toothed dolphins) is the best way to tell these two species apart, and both species often show their entire head on surfacing. The cookie-cutter scars on bottlenose dolphins also usually quickly repigment close to their background gray coloration, whereas adult rough-toothed dolphins, particularly on their bellies, have extensive white or pink blotches from these scars.

Habitat Use, Movements, and Abundance

Rough-toothed dolphins are found in tropical and warm temperate oceanic waters worldwide. In Hawaiian waters, this species has been sighted in deep offshore waters, around the Northwestern Hawaiian Islands, and around the main Hawaiian Islands, primarily in deeper waters. Overall in our study they are the fourth most frequently encountered odontocete, representing about 11 percent of all sightings, but relative abundance and habitat use vary dramatically among the main Hawaiian Islands. In our work off Kaua'i and Ni'ihau, rough-toothed dolphins are the most frequently encountered species of odontocete, representing more than a quarter of all sightings. Off Hawai'i Island or O'ahu they are encountered much less commonly, representing only about 6 percent of sightings off O'ahu and 10 percent of sightings off Hawai'i Island. Off Kaua'i and Ni'ihau they use relatively shallow waters, with more than half of our sightings in depths less than 1,000 m, while off Hawai'i Island less than 5 percent of our sightings of rough-toothed dolphins are in less than 1,000 m. The shallowest we've documented the species in Hawaiian waters is 280 m, and around the main Hawaiian Islands sighting rates increase with depth. Overall they are the most frequently encountered species of whale or dolphin in depths greater than about 3,000 m. Abundance estimates for all Hawaiian waters reflect this: in the 2010 NMFS survey, rough-toothed dolphins were estimated to have the highest abundance of any species of whale or dolphin, with an estimate of over 72,000 individuals!

Although rough-toothed dolphins are very abundant, strandings are relatively infrequent, representing less than 3 percent of all odontocete strandings documented in Hawai'i. They are one of the species that is known to mass strand, however. In June 1976, three different groups of rough-toothed dolphins stranded on Maui over a nineteen-day period, with a total of twenty-seven individuals stranding. Most of the animals died, either on the beach or after having being taken in for rehabilitation, but nine were released back to the ocean. Whether those individuals survived is unknown, but from genetic analyses of samples collected from some of the individuals, they appeared to be from an open-ocean population.

Despite their use of deep waters and a high overall abundance, evidence suggests that there are distinct island-associated resident populations off both Hawai'i Island and Kaua'i and Ni'ihau. Our photo-identification catalog of rough-toothed dolphins from Hawai'i includes over 1,100 distinctive individuals. Off the island of Hawai'i, where they spend most of their time in deep water, there is a very high resighting rate, with some individuals seen over spans of up to nine years. Using resightings from photo identification, we estimated that only about three hundred rough-toothed dolphins were using the west side of Hawai'i Island over a four-year period, compared to nearly two thousand individuals using the area around Kaua'i and Ni'ihau over a three-year period. We can also use the photos to examine movements, and we have documented movements of one group of twenty individuals from Hawai'i Island to Lāna'i. Other than this one group and for those living in the channel between Kaua'i and Ni'ihau, very few individuals have been documented moving among islands. Two individuals first documented off Kaua'i were seen less than a year later off Hawai'i Island, but they were not with any of the known Hawai'i Island resident rough-toothed dolphins. Several years later they were documented again off Kaua'i, mixed in with the Kaua'i residents, so their trip to Hawai'i Island did not represent a permanent move. One individual documented off O'ahu in 2003 was next photographed off Kaua'i more than eleven years later. Genetic studies of skin biopsy samples we've collected off Hawai'i Island, O'ahu, Kaua'i, and Ni'ihau also indicate that movements to or from Hawai'i Island are rare—the individuals off Hawai'i Island are genetically distinct from those elsewhere in the archipelago.

Since 2011, we have been using remotely deployed LIMPET satellite tags to examine movements of rough-toothed dolphins off Kaua'i and Ni'ihau, tagging fifteen different individuals between then and the end of 2015. Tagged individuals have circumnavigated Kaua'i and extensively use the channel between Kaua'i and Ni'ihau. Only one of the fifteen tagged individuals has moved away from those islands, visiting off the west side of O'ahu and returning.

Predators and Prey

Although we've seen killer whales only three times in Hawaiian waters, during one of those encounters off Kaua'i in 2011, they were seen chasing a rough-toothed dolphin. Although we did not witness the outcome of the attack, it is clear that killer whales are at least occasionally predators of rough-toothed dolphins. None of the individuals in our photo-identification catalog has healed wounds from large shark bites, but I suspect this reflects the likelihood that attacks by large sharks are almost always fatal rather than that such attacks don't occur.

Rough-toothed dolphins actively forage during the day, chasing small fish including flying fish, circling schools of small fish near the surface, and occasionally catching larger fish, primarily *mahi mahi.* When they catch large fish like *mahi mahi,* they typically break them up before eating, often peeling off parts to consume, and individuals may pass fish back and forth during consumption. Rough-toothed dolphins sometimes associate with floating debris such as nets and logs, and sometimes they congregate around FADs—Fish Aggregating Devices that have been deliberately anchored offshore to attract fish for fishermen—to take advantage of the fish that are associated with the debris or FADs. They will also try to scavenge fish from their larger cousins, the false killer whales. We've seen rough-toothed dolphins following closely behind a false killer whale that was carrying a fish and apparently trying to get bits of the fish away from the whale. Because rough-toothed dolphins feed on fish near the surface, we sometimes see birds associated with the groups, particularly great frigatebirds, which sometimes swoop down to the surface to try to catch flying fish scared out of the water by the dolphins.

Life History and Behavior

Female rough-toothed dolphins have been known to become pregnant by six years of age, although the typical age of sexual maturity for females is probably closer to eight to ten years. Males reach sexual maturity between five and ten years of age. Although our sightings of this species span throughout the year, the only months we've seen newborn calves are during the summer, May through August, suggesting that there may be a distinct breeding season for this species. Maximum life span is unknown, but it is probably in the range of thirty to fifty years.

While lone rough-toothed dolphins are occasionally observed, they are a fairly gregarious species, with an average group size of eleven individuals. Like other aspects of their behavior, group size varies among the islands: groups off Kaua'i and Ni'ihau average twice the size of those off Hawai'i Island. The largest group we've documented was off Kaua'i in September 2015 and had an estimated 140 individuals. While large groups are just temporary aggregations of individuals, analyses of association patterns show that rough-toothed dolphins do have preferred companions and strong associations, although these more stable groups are probably relatively small, with just a few individuals.

Rough-toothed dolphins are also seen with other species of marine mammals fairly regularly; 10 percent of our sightings have had other species present, not including the killer whale attack. The species we've seen them associating with most often are melon-headed whales and false killer whales, with the two species mixed together and intermingling. We've also seen them regularly associating with humpback whales during the winter, including seeing them bowriding on the whales, as well as interacting with short-finned pilot whales, bottlenose dolphins (once), and near both Cuvier's and Blainville's beaked whales.

A mother and newborn rough-toothed dolphin off Kaua'i, June 23, 2012. Photo by Daniel L. Webster.

Like other species, rough-toothed dolphins often have hitchhiking remoras and leap to try to get rid of them. This photo was taken off Kona, December 4, 2008. Photo by author.

When we first saw the net around the dorsal fin of this rough-toothed dolphin off Kaua'i in February 2011, we assumed it was entangled, but the dolphin then dropped the net off the fin and appeared to be playing with it. Such tendencies to investigate objects in their environment can have negative consequences, as individuals do get entangled occasionally. Photo by author.

As noted, rough-toothed dolphins are often seen foraging during the day, but from depth-transmitting satellite tags deployed on individuals off Kaua'i we know that they are diving more often and typically deeper at night than they do during the day. They are not particularly extreme divers—maximum dive durations recorded from tagged individuals range from about four to seven minutes. Compared to the other species for which we have diving data, they don't dive very deep, with a maximum dive depth noted of 399 m (although in human terms, 399 m, or 1,309′, is pretty extreme). Rough-toothed dolphins are very active at the surface, leaping out of the water fairly regularly and often multiple times in a row, although not typically as high out of the water as some of the smaller dolphins such as spinner and pantropical spotted dolphins.

Conservation

Twenty-three rough-toothed dolphins were captured in Hawaiian waters for aquaria in the 1960s and 1970s, most of them (fifteen) off of O'ahu. Whether these captures had any lasting impact on the population is unknown. There is only a single stock of rough-toothed dolphins currently recognized in Hawaiian waters, despite the evidence of a relatively small population that resides off the island of Hawai'i. Given that there are regular fishery interactions with this species off the island, this lack of recognition of a distinct population puts that population at greater risk. Rough-toothed dolphins have a reputation for taking bait and catch off the hooks of fishermen, and off the west coast of the island of Hawai'i, where there is a concentration of commercial and recreational fishing, these types of interactions may occur on a daily basis. There have been a number of reports of dolphins being hooked inadvertently when trying to take the bait or catch, although how often this occurs and what may be the typical outcome are unknown factors. Over the years, fishermen have tried many methods to deter dolphins from taking their catch, including leaving the area when rough-toothed dolphins show up, putting hooks into fish and throwing them to the dolphins, hoping this will discourage them if they eat the fish, and shooting at dolphins. How often fishermen shoot at rough-toothed dolphins is unknown, as is the outcome of such shootings when they happen. Such fishery interactions are reported for Hawai'i Island much more frequently than off Kaua'i, probably in part because there is relatively little fishing effort off Kaua'i in comparison, and the dolphins there may not have learned to associate fishing vessels with a potential easy lunch, as they have off Kona. From our observations, rough-toothed dolphins off Hawai'i Island behave as if they are shot at regularly, commonly avoiding the research boat when we get close. Off Kaua'i by comparison, they are often extremely interested in the boat, bowriding and generally easy to approach as long as they aren't actively foraging, even though we've collected dozens of biopsy samples and satellite tagged sixteen individuals there.

DOLPHINS OF THE GENUS *STENELLA*

There are three species within the genus *Stenella* in Hawaiian waters. These range in size from the spinner dolphin, reaching a maximum length of 2.1 m (~6′10″) and weighing up to 78 kilograms (kg) (~172 lbs), to the more robust striped dolphins, reaching a maximum length of 2.56 m (~8′5″) and a maximum weight of 160 kg (~352 lbs). Pantropical spotted dolphins are in between, reaching a maximum length of 2.4 m (~7′10″) and a maximum weight of 120 kg (~265 lbs). This variation in size corresponds to their varying habitats, with the small spinner dolphins being found in generally nearshore waters, the large striped dolphins in far offshore waters, and pantropical spotted dolphins inhabiting waters in between and overlapping with both of the other species. Although spinner dolphins are well known to the general public given the ease of seeing them close to shore and their traditional use of certain bays for resting, of the three species of *Stenella* in Hawaiian waters they are the least abundant. Both spinners and spotteds are sometimes referred to by fishermen in Hawai‘i as porpoises, a holdover from the use of this term by fishermen to describe dolphins in the eastern tropical Pacific.

An adult male spinner dolphin (top) and a pantropical spotted dolphin (bottom) off Kona, May 11, 2011. This spinner dolphin was documented offshore of Kona associated with spotted dolphins on a number of occasions in 2011 and 2012. Photo by Daniel L. Webster.

PANTROPICAL SPOTTED DOLPHINS *(Stenella attenuata)*

The first specimens of spotted dolphins from Hawaiian waters were collected off O'ahu in June 1901, with the skeletons deposited in the U.S. National Museum. When Ken Norris first came to Hawai'i in the mid-1960s to study dolphins, it wasn't the spinner dolphin that was his first focus of study but the pantropical spotted dolphin, or *kiko,* the Hawaiian word for "spot." In 1965 he tagged one spotted dolphin off the Wai'anae Coast of O'ahu, using a numbered plastic tag with a plastic pin through the dorsal fin. Three and a half years later he saw the

A heavily spotted pantropical spotted dolphin with a large remora attached, off Kona, October 30, 2009. On the back, at the leading edge of the dorsal fin, there is a distinctive rosette pattern formed from a cookie-cutter shark bite wound after it has repigmented. Photo by Daniel L. Webster.

tagged individual again and concluded that "the kikos are resident for sure, their chosen ground a portion of the Oahu coast." Ken worked with spotted dolphins for several years. Abandoning this pursuit, he noted in *The Porpoise Watcher* that it was "too expensive in time, money, and effort to work with the animals" and he began what was to become an extensive and thorough study of spinner dolphins off Hawai'i Island.

Despite being perhaps the most abundant dolphin in the waters immediately around the main Hawaiian Islands, very little research attention was concentrated on this species in Hawaiian waters over the next thirty years. When I first started working with whales and dolphins in the shallow waters off Maui in 1999, rather than working with the predictable spinner dolphins, which had been the subject of dozens of studies over the previous thirty years, I took the opportunity to work with other species in the area, including false killer whales, bottlenose dolphins, and pantropical spotted dolphins. In the first year, we encountered thirty-six groups of spotted dolphins in the basin between Maui, Lāna'i, and Kaho'olawe, ranging from as small as 10 individuals to as many as 150. We used suction-cup-attached tags with time-depth recorders to study their movements and behavior, deploying the tags on six different individuals. The dolphins moved into deeper water as the day progressed, and we discovered that they were diving deeper and longer at night than during the day. They also swam faster at night (the tags had a small paddle wheel that recorded swim speed), and right around sunset they performed a series of progressively shallower dives that tracked their prey—fish and squid associated with the "deep-scattering layer"—as it rose toward the surface. We also noted that the spotted dolphins off Maui and Lāna'i had relatively few scars from cookie-cutter shark bites compared to those off the island of Hawai'i, suggesting that those living in the shallow waters between Maui and Lāna'i might not venture into the deepwater channels between the islands where they would most likely encounter cookie-cutter sharks.

Identifying Features and Similar Species

Average length at birth is about 82 cm (2′8″), and when they are born, pantropical spotted dolphins actually lack spots. As they age they become progressively more and more spotted, with small white spots on the darker back and darker spots on the lighter belly. They have a dorsal cape that darkens somewhat as they age. The cape starts narrowly on the top of the head, becomes wider, and extends lower on the body in front of the dorsal fin before sweeping upward behind the dorsal fin. They have lighter-gray sides and a light-colored rostrum that also darkens with age, except for the tip, which becomes progressively whiter. They also have a faint but wide, dark stripe that extends from the eye to the mouthline. Although they are bitten by cookie-cutter sharks, when the wounds heal the skin seems to grow inward from the edges, and spots around the edge of the wound elongate and are pulled in to the center, giving them starburstlike patterns of white lines interspersed among the background of spots. Adult males average about 10 cm (~4″) larger than adult females, and in Hawaiian waters the adult males appear to

A pantropical spotted dolphin calf leaping in the waters off Kona, August 10, 2012. Photo by Daniel L. Webster.

Pantropical spotted dolphin adult male (top), adult female (middle), and juvenile (bottom). Adult males appear to be more heavily spotted than adult females and have a post-anal ventral keel. Illustrations by Uko Gorter.

be more heavily spotted than adult females, although this has not been quantified. Adult males also have a distinct ventral keel that is visible underwater or if they leap clear of the water.

Spotted and spinner dolphins are occasionally seen together and can easily be confused with each other, especially young animals. For juveniles, the most obvious way to tell them apart is the rostrum: in spotted dolphins it is uniformly light gray, while in spinner dolphins it is tricolored, with dark-gray lips and dark gray

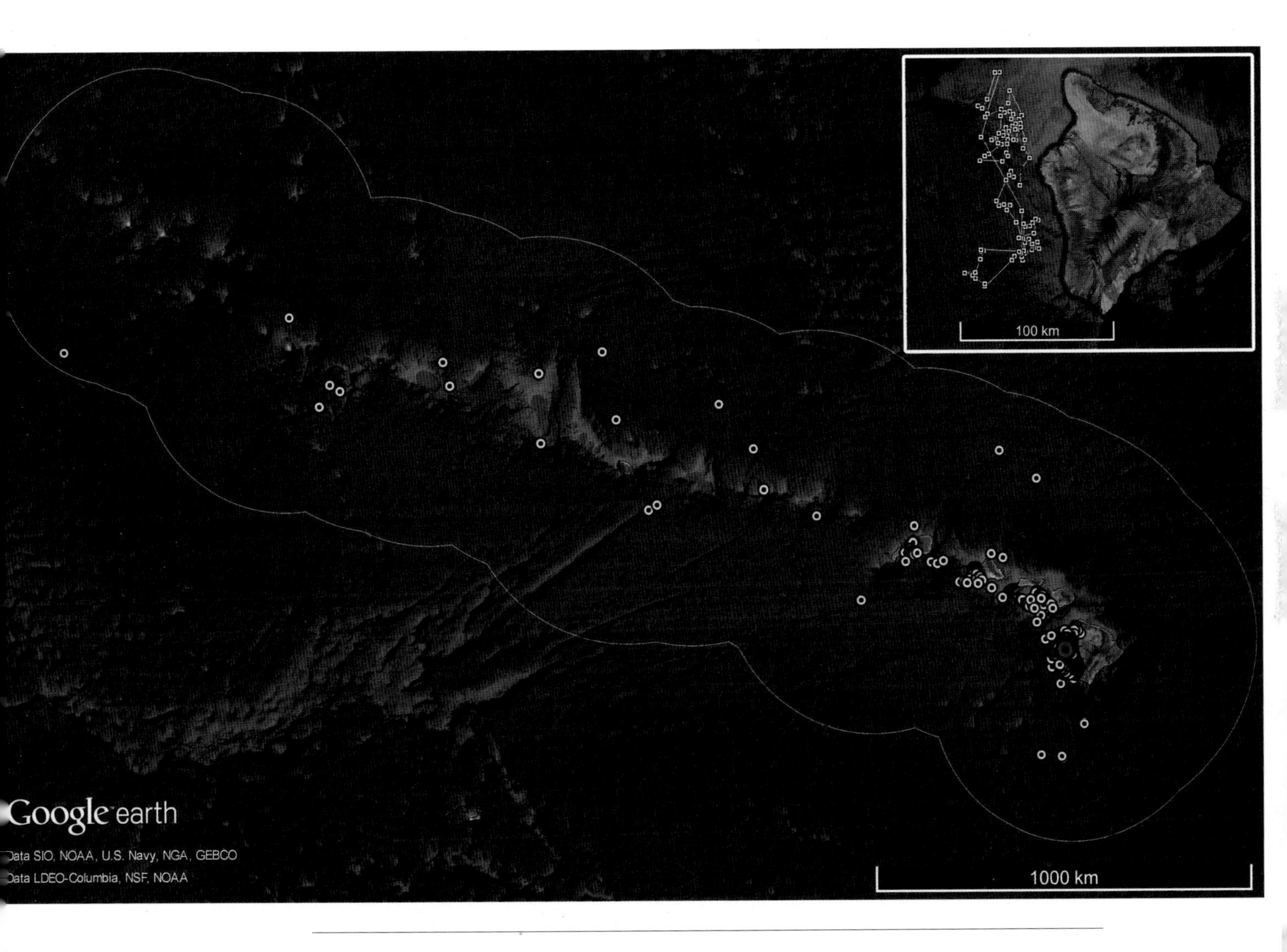

Pantropical spotted dolphin sighting records throughout Hawaiian waters, with the location of one satellite tagged individual indicated in red. The inset shows locations of an individual satellite tagged off Kona over an eleven-day period in April and May 2015.

above, a darker (almost black) tip, and white below. Spinner dolphins also have a very narrow mouth-to-eye stripe. The dark dorsal cape also dips much lower down for spotted dolphins than spinner dolphins. Some people have suggested the two species might hybridize in Hawaiian waters, leading to individuals with intermediate features, but there isn't any conclusive evidence for such hybrids.

Habitat Use, Movements, and Abundance

As their name implies, pantropical spotted dolphins are found throughout the tropics worldwide, but there are two subspecies: a coastal subspecies found only along the west coast of Central America and an offshore subspecies found elsewhere, including Hawai'i. Spotted dolphins have been seen throughout Hawaiian waters, but density is higher around the main Hawaiian Islands than in offshore waters or around the Northwestern Hawaiian Islands. Sighting rates peak in waters between 1,500 and 3,500 m deep; they are strongly associated with the island slopes.

Not a lot is known about the movements of individual spotted dolphins in Hawaiian waters. In 1980, a subadult female was captured off the west side of Hawai'i Island, radio-tagged, and tracked over a four-day period. During this time the dolphin meandered from the central to the northern Kona Coast, but no wide-ranging movements were documented. In April 2015, we remotely deployed a satellite tag on a spotted dolphin off Kona and tracked it for eleven days. Although it covered at least 700 km during that time, it remained off the Kona Coast for the entire period. The farthest it moved from where it was tagged was 75 km, and it stayed an average of only 36 km from the tagging location. We satellite tagged our second spotted dolphin in January 2016 off of Wai'anae and are still tracking it off O'ahu as I write this.

The best evidence for movements, or lack thereof, actually comes from genetic studies of spotted dolphins in Hawai'i. Skin biopsy samples we've collected from spotted dolphins around all the main Hawaiian Islands were used in genetic analyses as part of a PhD study by Sarah Courbis at Portland State University, and indicated that movements among island groups, such as from Maui Nui to O'ahu or Hawai'i Island, are extremely rare. Such movements happen so infrequently that three different reproductively isolated populations of spotted dolphins have been recently recognized: one off O'ahu, one off Maui Nui, and one off Hawai'i Island. Spotted dolphins off Kaua'i and Ni'ihau appear to be part of a widely ranging open-ocean population.

Spotted dolphins are one of the most abundant whales or dolphins in Hawaiian waters. The estimate from the 2010 NMFS survey for the open-ocean population was almost 56,000 individuals. The only species with larger estimates were rough-toothed dolphins and striped dolphins. Abundance estimates are not available for any of the island-associated populations, however. Relative abundance around the main Hawaiian Islands varies dramatically. Based on encounter rates and average group sizes, spotted dolphins appear to be the most abundant species of odontocete off O'ahu, around Maui Nui, and off Hawai'i Island, but they are only rarely sighted off Kaua'i and Ni'ihau. In our work off Kaua'i and Ni'ihau over the years, spending 161 days on the water between 2003 and the end of 2015, we've seen spotted dolphins on only seven different days, and four of those sightings were the same lone individual seen mixed in with a group of spinner dolphins.

Predators and Prey

Attacks by sharks on many species of whales or dolphins are often inferred by the bite marks on those that get away, and some species get away more often than others. With pantropical spotted dolphins, in 450 encounters among the islands, we've never seen an individual with a healed wound from a large shark bite. I suspect this is because most shark attacks on this species are fatal, rather than that they rarely get attacked. Pantropical spotted dolphins are one of the few species where an attack by a large shark, a tiger shark, has actually been observed in Hawaiian waters. During an aerial survey off Penguin Bank, researchers witnessed a large tiger shark attack and fatally wound a juvenile spotted dolphin that was lagging behind the group. We've been with spotted dolphins slowly moving through an area when all the animals suddenly leap clear of the water almost simultaneously and start porpoising rapidly away. In several of these cases, we've gone over to where the group was and spotted a large shark circling in the area. Clearly, spotted dolphins view large sharks as imminent threats and flee when they are seen. Pantropical spotted dolphins have been recovered from inside the stomachs of sharks elsewhere, and an attack by a smooth hammerhead shark off Brazil was also witnessed—likely successful, given the blood seen in the water.

Cookie-cutter sharks also take bites out of pantropical spotted dolphins, and while for most species of whales and dolphins in Hawaiian waters such bites are probably just a painful nuisance, in the case of spotted dolphins and their smaller cousins the spinner dolphins, some bites can be fatal. We've seen one spotted dolphin that had a bite into the abdominal cavity, with intestines visible through the wound. Although the dolphin did not seem in distress, it is hard to imagine the individual surviving long with that kind of wound. Spotted dolphins have also been observed being attacked by killer whales in Hawaiian waters. One attack was observed off Kona in 2003, and spotteds were observed avoiding killer whales in the same area in 2000.

We occasionally see spotted dolphins feeding during the day, chasing flying fish or other small fish, sometimes encircling and swimming through schools of bait fish, and with small fish in their mouths, but they don't seem to echolocate much during the day, suggesting that they aren't foraging much. Instead, they likely start feeding primarily around sunset and then feed throughout the night, as their deepwater prey rises up closer to the surface, where it is easier to access. During the day spotted dolphins stay relatively close to the surface, typically diving no deeper than about 50 m. But at night, one individual we tagged repeatedly dove to between 150 and 250 m all throughout the night, and the deepest dive we've recorded was to 342 m, also at night. Swimming speed at night is greater than during the day, suggesting that is when they are doing most of their hunting. Spotted dolphins feed on a wide diversity of small squid and fish, most associated with the deep-scattering layer.

Life History and Behavior

Much is known about the life history of pantropical spotted dolphins based on studies of individuals killed in the tuna purse seine fishery in the eastern tropical Pacific. Females reach sexual maturity at nine to eleven years of age, while males mature at about twelve to fifteen. Gestation lasts just less than a year, and one calf is born every two to three years. Maximum known lifespan is forty-six years, although it is possible that some individuals live longer. In Hawaiian

A newborn and an adult female pantropical spotted dolphin off Kona, August 18, 2012. The dorsal fin of the newborn is still partially bent over, indicating it was probably born within a few hours. Fetal folds, the pale vertical lines on the infant's side, are visible and will likely last several months. Photo by Daniel L. Webster.

waters, calves appear to be born over a long stretch of the year, at least from April through December, but there is a peak in the number of newborn spotted dolphins seen between July and October.

Pantropical spotted dolphins are typically found in fairly large groups, averaging about sixty individuals, and the largest group we've seen had an estimated four hundred individuals, seen off the island of Hawai'i. These groups are thought to be quite fluid, with individuals mixing and matching, although photo-identification studies to examine individual associations have not yet been undertaken. The groups can be spread out over large areas, sometimes even kilometers wide. Often within groups, spotted dolphins appear to be segregated

into subgroups by age and sex. We frequently see subgroups with multiple pairs of females and small calves, groups composed mainly of juveniles and subadults, and groups of very large and heavily spotted individuals that appear to be adult males. We've seen spotted dolphins nearby many other species of whales and dolphins, including mixed groups with spinner dolphins, rough-toothed dolphins, and bottlenose dolphins.

Spotted dolphins frequently approach boats to bowride, although this is more common for females and subadults than it is for adult males. They can be quite acrobatic, leaping two to three times their body length into the air; often when they do this, or when they repeatedly jump, they are trying to dislodge remoras (suckerfish). Young animals in particular seem to have great difficulty in getting rid of remoras.

Conservation

It is the association between spotted dolphins and *'ahi* that results in most of the interactions between humans and spotted dolphins, both in Hawaiian waters and in other parts of their range. In places like the eastern tropical Pacific, entire dolphin schools are regularly being encircled by purse seine nets in order to catch the associated fish. Since the late 1950s, over 6 million dolphins of various species have been killed in this fishery, including several million spotted dolphins. Modifications to fishing practices have dramatically reduced mortalities, but it is possible that the stress of being repeatedly captured and released, as well as probable separations of calves from mothers, are preventing the eastern tropical Pacific spotted dolphin population from recovering.

Spotted dolphins around the Hawaiian Islands are morphologically and genetically distinct from those in the eastern tropical Pacific, so the Hawaiian populations have not been impacted from the decades of purse seine fishing in the ETP. But in Hawaiian waters, spotted dolphin groups are targeted by smaller

‹‹‹

A pantropical spotted dolphin leaping with a remora attached, off Kona, August 28, 2011. Photo by Stacia D. Goecke.

vessels using a variety of hook and line methods to catch tuna that may be following the dolphin groups. We've observed such interactions off O'ahu, Maui Nui, and Kona. These include charter fishing vessels, typically trolling lines through groups of dolphins, as well as smaller vessels that move through a group of dolphins to the leading edge of the group as they travel and then stop and put lines over, using a form of chumming called *palu ahi* fishing or other methods of fishing as the dolphins move past the boat.

There are at least three ways that fishing around spotted dolphin groups could negatively impact individual dolphins. Pantropical spotted dolphins often approach boats and bowride or wake ride, and like other species that closely approach boats they are occasionally hit by vessel propellers, sometimes being killed

A pantropical spotted dolphin surfacing with visible injuries from a propeller wound, off Kona, May 23, 2003. Photo by Annie M. Gorgone.

as a result and sometimes living with the scars to show it. I suspect that dolphins getting hit and killed by vessel propellers are pretty rare events, but we have seen individuals with propeller strike wounds. A more important issue is the potential impact of one or more vessels, and the associated vessel noise, around groups of dolphins for extended periods, and how this may disrupt activity patterns. We know from tag data that spotted dolphins feed primarily at night, so like their cousins the spinner dolphins, their primary rest period is probably during the day. Almost two-thirds of the groups we encounter have no fishing vessels present, but for the other third, one or more vessels are fishing around or through the group of dolphins. During one encounter that lasted over six hours, there were fishing vessels with the group the entire time, and we recorded twenty-one different fishing vessels with the dolphins. An acoustic tag that we had deployed on one individual showed high levels of noise, both caused by the nearby vessels and by their echosounders. We don't see spotted dolphins avoiding boats—usually just the opposite—so I don't think they are easily disturbed by them. But having continuous engine and echosounder noise associated with boats following their group for extended periods definitely has the potential to disrupt normal behavioral patterns. Whether this is the case is unknown. Unlike spinner dolphins, which rest in nearshore bays where behavioral observations can easily be made from shore, the typical offshore habits of pantropical spotted dolphins make it more difficult to study their behavior in relation to boats.

Accidentally getting hooked or entangled in fishing gear is probably a much greater concern, and there are a number of reports of spotted dolphins getting caught in fishing gear in Hawai'i, with one individual reported being hooked and landed as early as 1962. The four populations of pantropical spotted dolphins in Hawaiian waters have been recently recognized by NMFS, and it is the three smaller island-associated populations that face the greatest conservation risks. It is not known how often individuals are hooked or how often this results in mortality. Because no abundance estimates are available for the three island-associated populations in Hawaiian waters, it is impossible to say whether the occasional hookings that are reported are affecting any of the populations.

A hooked subadult pantropical spotted dolphin off Kona, November 2, 2014. This individual was trailing about 5 m of line. Photo by Deron S. Verbeck/iamaquatic.com.

SPINNER DOLPHINS *(Stenella longirostris)*

With older and slower film cameras, it was possible to capture a small number of images as spinner dolphins would leap out of the water and perform their characteristic spin, and there has been much debate as to the purpose of this behavior. Individuals of all ages spin, and when spinner dolphins have been kept in captivity, they also spin. When higher-resolution digital cameras capable of taking multiple frames per second became available, and with increases in both resolution and the number of frames that could be captured, it became clear when examining photos of spinner dolphins spinning that the vast majority of these iconic leaps are actually attempts to get rid of remoras, fish with a suckerlike apparatus that has backward-facing barbs that dig into the skin. Many spinners, particularly juveniles, suffer from what I call persistent remora damage, a large raw patch just below the dorsal fin that forms when remoras remain attached for too long or are too vigorous in their attempts to remain closely bonded with their hosts. Remoras are often thought of as benign hitchhikers, but at least in the case of some spinner dolphins, as well as pantropical spotted dolphins and less often bottlenose dolphins, it is clear that the damage they do is both irritating to the dolphins and potentially causes a wound that could become infected.

Given their nearshore habits, spinner dolphins were likely well known to the ancestors of today's Native Hawaiians. They were probably seen by the first European explorers, including Captain James Cook and his crew when they were in Kealakekua Bay in 1779, an area regularly used today by resting spinner dolphins. Much of what we know about spinner dolphins in Hawaiian waters comes from a long-term study by Ken Norris, his colleagues, and his graduate students who worked off Hawai'i Island, mainly in Kealakekua Bay, over a twenty-five-year span starting in the late 1960s. Ken's pioneering work provided an in-depth view into the lives of spinner dolphins, both above and below the water. It also

A spinner dolphin spinning in the waters off Kona, July 18, 2010. Subsequent frames in this series showed a small remora on the belly between the flippers. Photo © Dan J. McSweeney/Wild Whale Research Foundation.

A spinner dolphin leaping in an attempt to remove two remoras, off Kaua'i, October 9, 2014.
Photo by Daniel L. Webster.

spawned many other efforts to learn about this species in Hawaiian waters, including studies in the Northwestern Hawaiian Islands—spinners are the only species of whale or dolphin that has been studied extensively there in addition to the main Hawaiian Islands. Because of Norris's work and the many others who followed him, my attention to spinner dolphins has always been somewhat limited compared to the other species of odontocetes in Hawai'i. Yet, they are hard to miss: working out of Honokōhau Harbor on the west side of Hawai'i Island every year since 2002, we would see them frequently as we headed in and out of the harbor, as the area immediately outside the harbor is a traditional resting area for spinners. It is this tendency of spinner dolphins in Hawai'i to spend their days resting near shore that has made them a comparatively easy species to find and study, but it has also led to serious concerns about their conservation and management.

In the Hawaiian language, spinners and other dolphins are known as *nai'a*. They are often referred to as "Hawaiian" spinner dolphins, but in fact those individuals in Hawai'i are part of a widely distributed pantropical subspecies, the Gray's spinner dolphin (*Stenella longirostris longirostris*). More is known about Gray's spinner dolphins in Hawaiian waters than any other population of this subspecies in the world.

Identifying Features and Similar Species

Spinner dolphins are the smallest and most streamlined cetacean in Hawaiian waters. They are about 77 cm (2′6″) at birth, and the longest animal measured in Hawai'i was 2.1 m (6′10), an adult male that stranded on Maui in March 2008. Spinner dolphins have the longest and slenderest rostrum of any species in Hawai'i. They are effectively tricolored, with a dark-gray cape, lighter-gray sides, flippers, and tail stock, and a white belly with the white extending partway up the sides. They have a thin dark stripe that extends from the eye to the rostrum and a wider dark-gray stripe that extends from the eye to the flipper. They also have a dark (almost black) tip on both the upper and lower jaws, and the upper jaw appears to become darker with age. When engaged in vigorous activities, the belly can turn quite pink. As adults they are sexually dimorphic, although less so than some other subspecies of spinner dolphins. Older males acquire a subtle

The Gray's spinner dolphin subspecies found in Hawaiian waters are slightly sexually dimorphic, with the adult males (top) being larger and having a dorsal fin that is more triangular. Illustrations by Uko Gorter.

ventral keel, and the dorsal fin of some older males appears virtually triangular, sometimes almost canted forward, as if it were placed on backward.

Spinner dolphins are most likely confused with pantropical spotted dolphins, and the two species sometimes associate, making it even harder to determine what kind of school of dolphins one is witnessing. In general, the dorsal fin of a spinner dolphin is less falcate (curved back) than that of a spotted. For juveniles, the tricolored rostrum is the easiest way to distinguish them (see the section on pantropical spotted dolphins).

A group of spinner dolphins surfacing off Lāna'i, December 13, 2012. Photo by Annie B. Douglas.

Habitat Use, Movements, and Abundance

Gray's spinner dolphins are distributed throughout the tropics worldwide, in both the open ocean and near oceanic islands. While they are widely spread throughout the eastern tropical Pacific, sightings offshore of the main Hawaiian Islands are extremely rare. While in theory there should be an open-ocean population of spinner dolphins in offshore Hawaiian waters, they seem to be largely absent. In the 2002 and 2010 NMFS surveys, covering all Hawaiian waters, there was only a single sighting of spinners far offshore and another closer to shore north

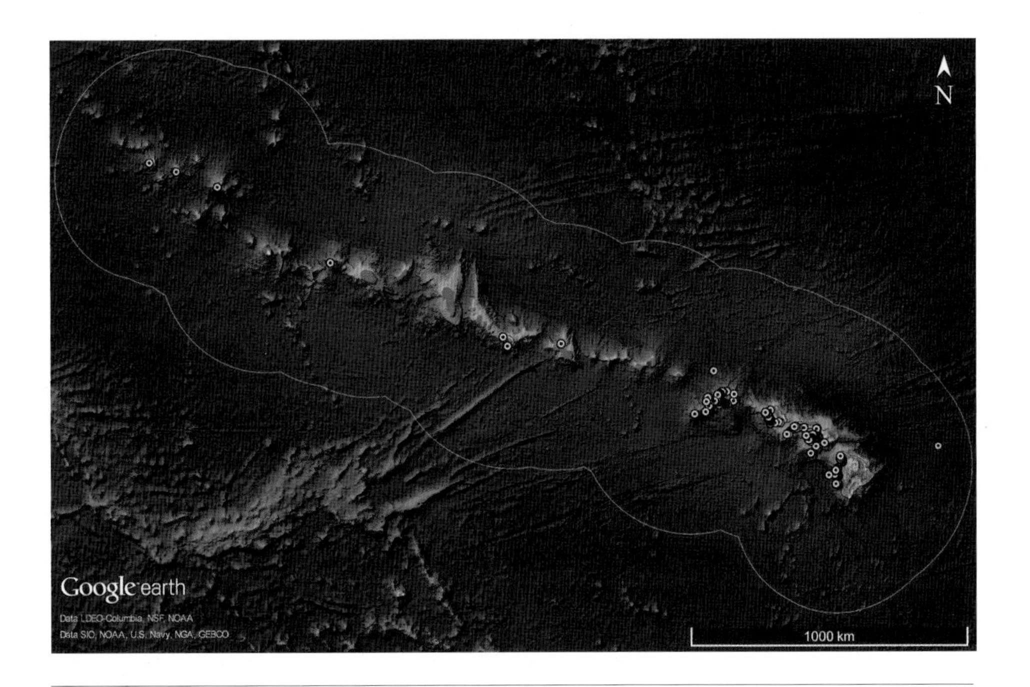

Spinner dolphin sighting records in Hawaiian waters, from NMFS research cruises and our boat-based fieldwork from 2000 through mid-2015. Despite extensive survey efforts in offshore waters, all but two sightings are associated with the islands.

of Kaua'i; all the remaining sightings were among the main and Northwestern Hawaiian Islands. They are found around all the main Hawaiian Islands, as well as around at least four in the Northwestern Hawaiian Islands: Kure, Midway, Pearl and Hermes, and French Frigate Shoals. There are many similar insular populations of Gray's spinner dolphins throughout the tropics, including in French Polynesia, in the Marquesas, off Palmyra Atoll, in the Northern Mariana Islands, and around Fernando de Noronha, an island off the coast of Brazil, to name a few.

Although we've seen spinners more than two hundred times, a large number of those sightings were right off the entrance of Honokōhau Harbor in Kona and Kīkīaola Small Boat Harbor on Kaua'i—both traditional resting areas for spinners. Spinner dolphins are the only species of cetacean in Hawaiian waters that shows such predictability in where they can be found, primarily because their requirements for good daytime resting habitat seem to be fairly strict, at least off most of the islands. Most of the areas where spinners rest during the day are shallow, sandy-bottom areas; the lighter-colored sand is thought to help spinners detect sharks that may approach. At Kure and Midway, spinners rest inside the lagoons of the coral atolls where no other cetacean species have been documented. Around each of the main Hawaiian Islands, there are usually a handful of particular bays or other nearshore areas where spinners predictably rest. In the main islands, they aren't necessarily found in any one area every day or even all day; when there are multiple suitable areas available for them, they may move among them during the day.

Sighting rates of spinner dolphins are highest in less than 500 m of water and drop off dramatically after that. We've never seen what I would consider to be offshore or pelagic spinners in Hawai'i. We've only seen them in depths greater than 1,000 m on eight occasions, and seven of those were sightings of one to three individuals hanging out with large groups of pantropical spotted dolphins. Only one of the deepwater groups, seven individuals, was alone. But they were heading offshore in water about 1,790 m deep when we encountered them in the late afternoon; presumably they were on their way out to get an early start on feeding.

Spinner dolphins tend to be homebodies, although there are exceptions. Spinners have been documented moving between Kure and Midway Atolls, a distance of 90 km (56 mi), and from Pearl and Hermes to Midway, 155 km (96 mi). However, such movements appear to be rare and have involved subgroups moving between the atolls. They regularly move back and forth between Kaua'i and Ni'ihau and among the islands in the Maui Nui Basin, but movements of spinners across the deeperwater channels in the main Hawaiian Islands are rare. One individual has been documented moving from Ni'ihau to O'ahu, two from O'ahu to Hawai'i Island, and six from Maui Nui to Hawai'i Island. Despite such exceptions to the rule, the rates of movements are low enough that spinners in different island areas are genetically differentiated, and five different island resident populations are recognized—Kure and Midway, Pearl and Hermes, Kaua'i and Ni'ihau, O'ahu and Maui Nui, and Hawai'i Island—as well as a pelagic population. There is no abundance estimate for the pelagic population, but there are estimates for most of the island-resident populations, based on mark-recapture analyses of photo-identification data: Kure and Midway—about 260; Kaua'i and Ni'ihau—about 600; O'ahu and Maui Nui—300 to 400; and Hawai'i Island—600 to 700 individuals.

Predators and Prey

Large sharks are probably the most significant predators of spinner dolphins in Hawaiian waters, and most attacks are likely fatal—we've never seen a spinner dolphin with a healed wound or scar from a large shark bite. I know of only one animal in the main Hawaiian Islands that has been photographed with a wound from a shark bite that wasn't a cookie-cutter shark, and from the size of the bite it was probably a medium-sized shark, something much smaller than an adult tiger shark. Given their small size, the smallest of the Hawaiian odontocetes, and thin blubber layers, even cookie-cutter shark bites are occasionally fatal: a calf was found dying off Kona in 2014 with a bite wound into the abdominal cavity. In theory, killer whales might also prey on spinners in Hawaiian

A female spinner dolphin calf with a fatal bite wound from a cookie-cutter shark, off Kona, October 1, 2014. Photo by Deron S. Verbeck/ iamaquatic.com.

waters, but the two species probably do not overlap much, given the nearshore habits of spinners in Hawaiʻi. Spinner dolphins in Hawaiʻi do all of their feeding at night, when they feed on a variety of small squid and fish associated with the deep-scattering layer. Myctophids, various types of lanternfishes, make up most of the prey, and they've also been documented feeding on small shrimp. Their prey not only come up closer to the surface at night but also come up the island slopes, approaching closer to shore. Spinners move from their nearshore daytime resting areas into offshore waters in the late afternoon and start feeding near sunset, following their prey as they move closer to shore in the middle of the night. Then, assuming they are still hungry, spinners move back offshore early in the morning following the prey out, before heading into their daytime resting areas again.

Life History and Behavior

Although they are the most frequently reported species of whale or dolphin stranding in Hawaiʻi, there is surprisingly little work that has been done on the life history of spinner dolphins here. Most of what is known comes from spinners killed in the tuna purse seine fishery in the eastern tropical Pacific. Females typically give birth at about seven years of age after a gestation of approximately eleven months. On average, they have one calf every three years, and females may live to about thirty years of age. We've seen newborn spinners in ten months of the year, suggesting that calving can occur at any time of year.

Spinner dolphins in Hawaiʻi are fairly social—the average group size of our encounters is thirty individuals, and the largest group we've seen had an estimated three hundred individuals. How stable these groups are varies dramatically within Hawaiian waters. At Midway and Kure, groups tend to be very stable, the same aggregation of individuals spending their time together most days. In the main Hawaiian Islands, spinners are thought to mix and match on a daily basis. Resting groups form in the morning, but at night, when spinners are offshore foraging, they may interact with many others and form new aggregations the next day. Spinners do not tend to associate with other species very often. The species we've seen them with the most is pantropical spotted dolphins. Sometimes we've seen small numbers of spotteds mixed in with larger groups

A spinner dolphin with a spotted dolphin calf, offshore of Kona, August 29, 2011. When we first saw these two together, we thought the calf might be a hybrid, but a later examination of photos of the spinner dolphin in the air revealed it was an adult male, suggesting it may have kidnapped the calf from a spotted dolphin mother. Photo © Dan J. McSweeney/Wild Whale Research Foundation.

of spinners in their nearshore resting areas, and other times we've seen a couple spinners associating with much larger groups of spotted dolphins in offshore waters. Occasionally these types of associations can last for extended periods. We observed one distinctive male spinner dolphin with a group of spotted dolphins offshore of Kona two years in a row, and one spotted dolphin male was seen with spinners off Kaua'i over a six-year period.

Conservation

Over the years, spinner dolphins in Hawai'i have faced a variety of threats. In the 1960s and 1970s, at least eighty-five spinner dolphins were captured in Hawaiian waters, primarily for captive display, but seven of them were either released or escaped. The last captive spinner dolphin at Sea Life Park was released back to the wild in 1983.

Spinner dolphins often encounter marine debris in the environment, including discarded bits of netting, lines, packing bands, plastic bags, and other garbage either thrown overboard or washed offshore. They are curious animals and will investigate and play with such debris, sometimes with unfortunate consequences. A number of animals have been reported or photographed with various types of debris wrapped around the head or stuck in the jaw, limiting movements and breathing, and this type of entanglement may end up killing individuals.

One spinner dolphin that was found dead on O'ahu died of toxoplasmosis, a disease caused by a terrestrial parasite that is transmitted, among other ways, via cats and cat droppings. There is a considerable feral cat population in Hawai'i, with many cat "colonies" being provisioned by well-meaning people, and these cat colonies may be a source of toxoplasmosis introduction into the marine environment. Of all the cetaceans, spinner dolphins are likely the species most impacted by the presence of toxoplasmosis, given their daytime use of nearshore areas.

The biggest conservation concern for spinner dolphins in Hawaiian waters, however, comes not from marine debris, fisheries, or naval sonar but from the intense interest in them by the general public. There are dozens of boat tour operations that focus on spinners and put people in the water to swim with them. Spinners also attract many people who swim out from shore to swim and interact with them. Because spinners do all of their feeding at night and all of their resting

Spinner dolphins bowriding on the research vessel off O'ahu, October 22, 2010. Photo by Jonas P. Webster

during the day, particularly in areas that may allow them to easily detect approaching predators, the issue is whether exposure to vessel traffic and swimmers may disrupt their resting patterns or cause them to leave the relative safety of their traditional resting areas. A recent study off Kona showed that spinner dolphins were exposed to humans, boaters, and/or swimmers within 100 yards of them about 82 percent of their time during the day. It is clear, however, that not all swimmers or tour boats are the same. Many approach groups slowly and carefully or allow individuals to come to them, while others may do so quickly and pursue individuals in the water. It is likely that the latter have greater impacts. Disrupting resting patterns may have long-term but difficult to quantify implications; dolphins may increase how much energy they expend, and they may not be able to make up for this without compromising some other aspect of their lives, such as reproduction. If spinner dolphins leave relatively safe resting areas, there may be more immediate and drastic consequences: when moving along the coast looking for respite from swimmers or vessel traffic, they may be more vulnerable to being attacked and killed by large sharks, which are common off of Hawai'i's shorelines.

STRIPED DOLPHINS *(Stenella coeruleoalba)*

Striped dolphins are an offshore species found from tropical to warm temperate waters throughout the world's oceans. In the eastern tropical Pacific they are referred to by fishermen as "streakers," reflecting their speed and tendency to take long leaps out of the water as they swim rapidly away. They were first recorded in Hawai'i in 1958, when an individual was captured in the Ala Wai Canal, presumably on its way to strand, given the oceanic habits of the species.

Although they are one of the most abundant oceanic dolphins in Hawaiian waters, sadly the first glimpse most boaters in Hawai'i may have of a school of striped dolphins will be of them leaping rapidly away. This species seems to actively avoid boats more so than any other in Hawaiian waters. Boat avoidance and their offshore habits mean that they are rarely seen around the main Hawaiian Islands, and they are a poorly known species.

Identifying Features and Similar Species

Striped dolphins are born at about 90–100 cm (2′11″–3′3″) long and reach maximum sizes of about 2.6 m (8′6″), and they are generally more robust than the other two closely related species in Hawai'i, pantropical spotted dolphins and spinner dolphins. All three species have a distinct beak or rostrum that is sharply demarcated from the head, and of the three striped dolphins have the shortest rostrum. They also have one of the most striking color patterns of any species of dolphin in Hawaiian waters. They have the typical countershading, with a dark-gray back, lighter-gray sides, and a white belly, but they also have bold dark stripes running from the eye to the flipper and from the eye all the way back to the anal area. Most individuals also have a lighter blaze that sweeps upward in the darker area below the dorsal fin; this latter feature is something also seen in bottlenose dolphins, albeit it is less obvious for bottlenose given their paler overall patterns. Given how frequently striped dolphins leap out of the water—in our case during almost every encounter we've had—it can be quite easy to see the stripes and dorsal blaze. Although striped dolphins in Hawai'i do show signs of bites from cookie-cutter sharks, it appears as if the wounds repigment quickly and do not substantially change their color pattern, unlike some species of delphinids.

The low arching leaps of a group of striped dolphins as they avoid the boat off Hawai'i Island, December 8, 2008. Photo by Russel D. Andrews.

There is one other species of medium-sized delphinid with which striped dolphins might be confused, as it also typically avoids vessels, usually leaping out of the water to do so: the Fraser's dolphin. Fraser's dolphins also have a bold lateral stripe, but their dorsal fin appears disproportionately small; they also have only a nubbin of a beak, and Fraser's dolphins are extremely uncommon around the main Hawaiian Islands. If striped dolphins aren't leaping, they are most likely to be confused with fast-moving pantropical spotted dolphins. They also associate on occasion with spotted dolphins, making it particularly difficult to confirm the species when quick glimpses may be of either species. When fast swimming but not leaping clear of the water, they may throw up a characteristic "rooster tail" splash.

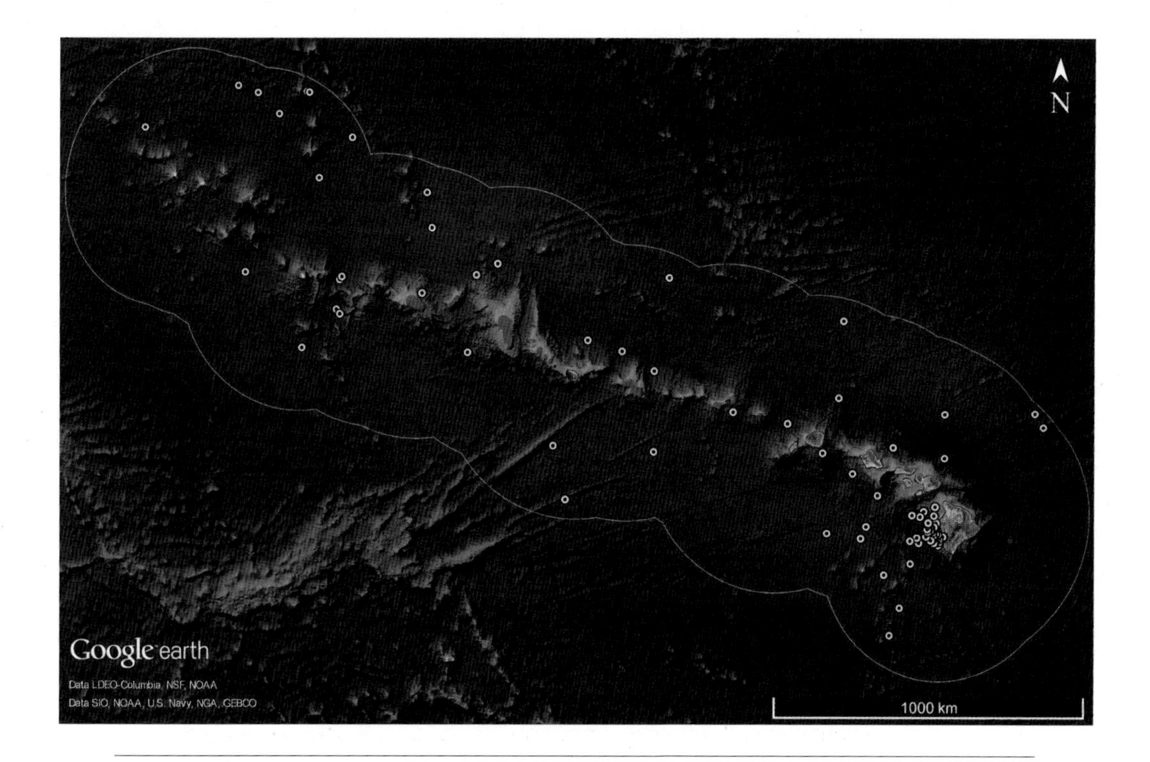

Striped dolphin sighting records in Hawaiian waters.

A pair of striped dolphins leaping off Kona, December 13, 2010. The color pattern of the individual in the foreground is muted and it lacks the dorsal blaze. Several well-healed scars from cookie-cutter shark bites are visible below the dorsal fin. Photo by author.

Habitat Use, Movements, and Abundance

Striped dolphins have been recorded throughout Hawaiian waters in all seasons, with numerous sightings in offshore waters, a couple sightings off the Northwestern Hawaiian Islands, and sightings off the main Hawaiian Islands. In our work, while we've seen them twice in depths shallower than 1,000 m, sighting rates increase dramatically in depths greater than 3,500 m; Risso's dolphins are the only species of dolphin in Hawai'i that seems to consistently use deeper waters than striped dolphins. In our work we've seen striped dolphins on thirty-five occasions through January 2016. They represent less than 2 percent of all of our odontocete sightings, but in waters deeper than 3,000 m they are the third most frequently encountered species, accounting for over 12 percent of odontocete sightings. Thirty-four of our sightings were off Hawai'i Island, with the one other sighting off Kaua'i, but they've been reported in deep waters off all the main Hawaiian Islands. Because of their avoidance behavior and the difficulty of getting good photos of many individuals, we've not yet compiled a photo-identification catalog of striped dolphins, nor have we been able to approach close enough to satellite tag individuals; thus nothing is known of their movements in Hawaiian waters. Striped dolphins are one of the four species of oceanic dolphins that likely do not have island-associated populations, the other three being Fraser's dolphins, Risso's dolphins, and killer whales.

Although not seen frequently around the main Hawaiian Islands, in offshore waters striped dolphins were the second most abundant of the oceanic dolphins in the 2010 NMFS survey, with abundance estimated at over 61,000 individuals.

Life History and Behavior

Despite their abundance and wide distribution, studies of free-ranging striped dolphins are rare anywhere in the world. Most of what is known about this species comes from studies of dead stranded animals or animals killed in fisheries. The oldest documented age is fifty-eight, and they become sexually mature from five to thirteen years for females and seven to fifteen years for males.

Striped dolphins are typically seen in fairly large groups. The mean group size in our studies is 28 individuals, and the largest group we've seen was of 110 individuals. We've never seen a lone striped dolphin, and while in one case we saw just a pair of individuals, given their typical behavior (avoiding the boat) and our short encounter durations (our average is only four minutes), it is likely that another larger group was somewhere nearby. From large-vessel surveys in offshore Hawaiian waters, the mean group size was about 53 individuals, likely a more accurate representation of their usual group sizes in Hawaiian waters. Virtually nothing is known of their social organization or behavior in Hawaiian waters or, to a large degree, elsewhere. In Hawai'i they have relatively few fresh cookie-cutter shark bite wounds, which suggests they probably do not spend a lot of time deep in the water column in Hawaiian waters, although elsewhere they are thought to dive from 200 to 700 m deep.

RISSO'S DOLPHINS *(Grampus griseus)*

More than twenty years ago, while out in a small boat in the Sea of Cortez, Mexico, I had a group of about seventy-five Risso's dolphins all bowriding on the boat, while we were moving slowly in calm seas. Sadly, I have not experienced such an event again; when we see Risso's dolphins in Hawaiian waters, they tend to avoid the boat. They don't flee like striped or Fraser's dolphins, they just slowly turn away or dive early, making it very difficult to get good photos. In April 2015,

The color of a Risso's dolphin varies with age. Illustrations by Uko Gorter.

The front half of the body of a Risso's dolphin becomes progessively whiter with age, apparently due to the accumulation of scars from interacting with other dolphins. This photo is from one of our close encounters with Risso's dolphins offshore of Kona, April 21, 2015. Photo by Annie M. Gorgone.

we had our tenth encounter with Risso's dolphins in Hawaiian waters. Learning from previous encounters, we approached them slowly, avoiding any changes in speed or dramatic changes in direction, just slowly moving in the same direction as the dispersed group of individuals. After a while they seemed to get used to our presence, and occasionally individuals would surface off to the side of the boat. During one of these events, we slowly moved the boat closer and were able to get close enough to deploy a satellite tag—the first satellite tag deployed on a Risso's dolphin in the central Pacific. Individual Risso's dolphins are extremely distinctive, extensively covered with tooth rakes on their bodies, and we are slowly building up enough good encounters to set up a photo-identification catalog of this species in Hawaiian waters.

Risso's dolphins are found worldwide in temperate and tropical areas and are the fourth -largest delphinid in Hawaiian waters, after killer whales, short-finned pilot whales, and false killer whales. Risso's dolphins were first documented in Hawai'i in 1977, when an individual stranded on the coast below Wailuku, Maui. They are born at about 1.1 to 1.5 m (3′7″ to 4′11″) in length and reach maximum lengths of about 4 m (13′1″). They have a rounded head, like the blackfish, but with a distinct vertical crease on the front, although the crease is very difficult to see in the field. Their dorsal fin is dark gray, but their body coloration changes with age. Perhaps more so than any other delphinid, Risso's dolphins accumulate distinctive white scars with age. For older adults, most of the front half of their body is covered with extensive white scarring caused by a combination of apparently fighting with other Risso's dolphins and, at least on the head, from scars inflicted by the primary prey, squid. From a distance, older animals appear to have an almost entirely white body and a dark fin. Their dorsal fin is larger in proportion to their body than any other species of whale or dolphin in Hawaiian waters, with the exception of adult male killer whales. As a result, Risso's dolphins are probably confused with killer whales more so than other species. If it is possible to see coloration, the generally white body of adults combined with the large dorsal fin makes this species relatively easy to discriminate from others.

Around the main Hawaiian Islands, Risso's dolphins are only infrequently seen, as they primarily use very deep waters. We've had only ten sightings of this species through January 2016, all off Hawai'i Island, which translates to just one sighting every 676 hours on the water. While we've had one sighting in

An adult Risso's dolphin offshore of Kona, April 21, 2015, just prior to being tagged with a LIMPET satellite tag. Photo by author.

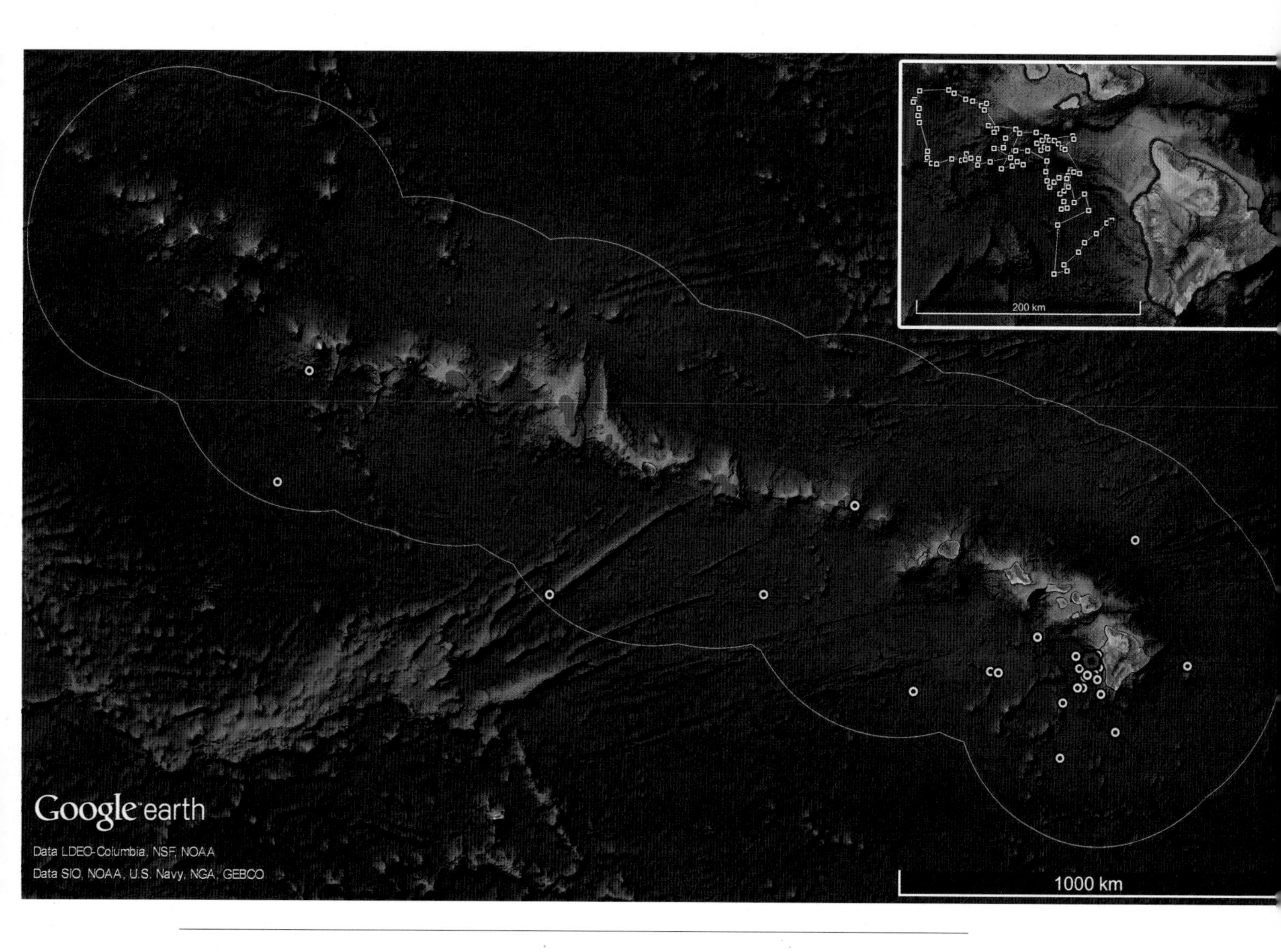

Risso's dolphin sighting records (yellow circles) in Hawaiian waters. One individual was satellite tagged April 21, 2015, with the tagging location indicated in red. The inset shows the fourteen-day track of the tagged individual.

just over 1,300 m depth, that was a lone individual associating with a group of short-finned pilot whales. Otherwise our sightings are in deep water, and sighting rates increase in depths greater than 3,000 m, with highest rates in depths over 4,500 m. We obtained two weeks of movement data from the individual we satellite tagged in April 2015, and during that time the dolphin meandered in deep water to the west of Kona and west of Kahoʻolawe and Lānaʻi. Combined with their deepwater habits and the infrequent sightings, it seems unlikely there would be any sort of resident population of Risso's dolphins around the islands.

A juvenile Risso's dolphin that was entangled and died in a shallow-set longline in international waters south of the main Hawaiian Islands, January 17, 2006. Photo courtesy of Pacific Islands Region Observer Program, NMFS.

Groups in Hawaiian waters tend to be relatively small, at least off the island of Hawai'i. Our average group size is only four individuals, and the largest group we've documented was twenty-five individuals. In far offshore waters, groups tend to be larger. The ten sightings of this species during the NMFS 2010 survey ranged from nine to about fifty individuals, with an average of twenty-two individuals. The abundance estimate from the NMFS 2010 survey was of more than 11,000 individuals, suggesting they are fairly common in offshore waters.

Given their offshore habits in Hawai'i, Risso's dolphins likely have few interactions with humans around the main Hawaiian Islands, but they are the species that is reported most often hooked in the Hawai'i-based swordfish longline fishery, primarily getting hooked as they try to take bait off the lines. Most of the Risso's injured or killed in that fishery are to the north of Hawaiian waters, and bycatch rates within Hawaiian waters are low relative to what the population is thought to be able to sustain. Elsewhere they have been intentionally killed, sometimes because of their tendency to take fish or bait from lines, and there is also a drive fishery for Risso's dolphins in Japan, with several hundred killed each year.

FRASER'S DOLPHINS *(Lagenodelphis hosei)*

It was over sixty years between when a dolphin skull was found on a beach in Sarawak, in what is now Malaysia, and when it was described as a new species by F. C. Fraser, whose name it now bears, in 1956. It wasn't until 1971 that this new species was seen alive, when several animals were caught and killed in tuna purse seine nets in the eastern tropical Pacific. The Fraser's dolphin remains a poorly known species, primarily because of its distribution in offshore tropical waters. Unlike many of the species discussed here, there is no evidence that Fraser's dolphins ever associate with the slopes of oceanic islands, including in Hawai'i. The only regions where they seem to regularly come close to shore are in deepwater

As adults, male (top) and female (bottom) Fraser's dolphins can be easily distinguished. Illustration by Uko Gorter.

A group of leaping Fraser's dolphins off Kona, April 21, 2015, showing sex differences in pigmentation patterns. The upper-left individual with the faint lateral stripe is likely an adult female, based on body size and the lack of a ventral keel, while the indivdual below is an adult male, with a bold black facial mask and lateral stripe. Photo by Annie M. Gorgone.

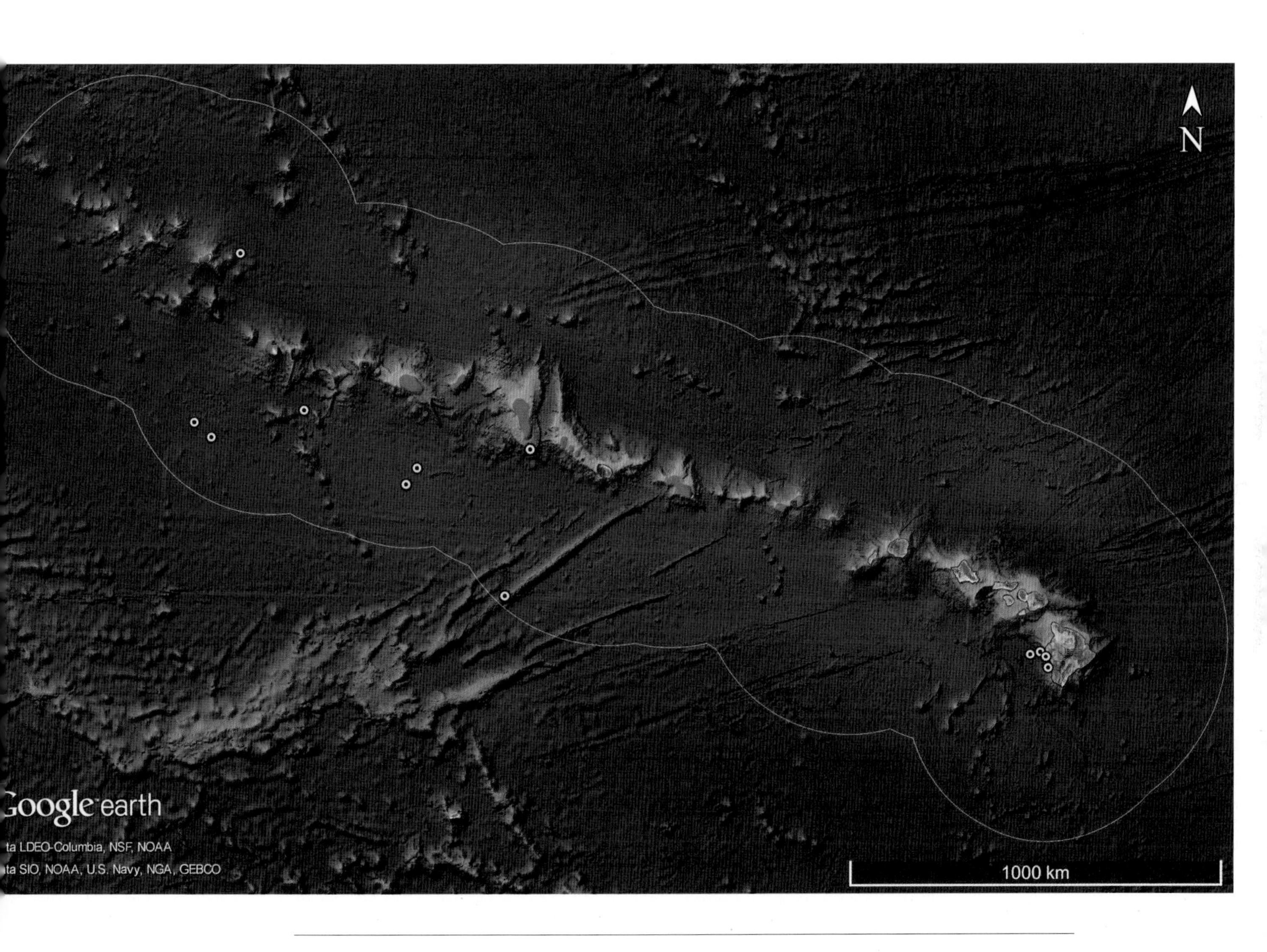

Fraser's dolphin sighting records in Hawaiian waters, from NMFS surveys, the IWC POWER (Pacific Ocean Whale and Ecosystem Research) cruise, and ongoing Cascadia research from 2000 through the end of 2015.

Subadult Fraser's dolphins off Kona, April 26, 2015, showing the relatively small dorsal fin and flippers and short rostrum diagnostic of this species. Photo by author.

areas in the Philippines, Indonesia, and the Lesser Antilles. Their body proportions are very distinctive. They are the most robust looking of the small delphinids in Hawai'i. They have a very small dorsal fin, much smaller in proportion to the body than any of the other Hawaiian delphinids, and a small beak, smaller than a bottlenose dolphin. Females and calves are fairly drab in coloration, with a gray cape and a lighter belly, but males are very distinctive, with a wide black stripe running from the beak all the way back to the genital area, as well as a narrower eye-to-flipper stripe.

Fraser's dolphins were first documented in Hawaiian waters during the 2002 NMFS survey, with two sightings of large groups far to the west of the main Hawaiian Islands. There was a stranding of a single Fraser's dolphin on Kaua'i in 2004, and there were four sightings of large groups of the dolphins in the 2010 NMFS survey, far to the west of the main Hawaiian Islands. We've seen this species only four times in our surveys, in April 2008, May 2012, and twice in April 2015, with groups of 75 to 110 individuals. In two of the four cases, they were seen associated with melon-headed whales, albeit not in mixed groups. All of our sightings were in relatively deep water (1,515 m to 4,600 m), and in all cases the groups avoided the boat, fleeing like striped dolphins tend to do in Hawaiian waters. I suspect that Fraser's dolphins are found in deeper waters around the main Hawaiian Islands more often than is reflected by sightings. Because they tend to avoid boats, unless they are observed through binoculars they are likely to be mistaken for a more common species, especially if they are often found on the periphery of large groups of melon-headed whales. Despite the rarity of sightings, because of the large average group sizes (almost three hundred) in the NMFS surveys, the abundance estimate for Fraser's in Hawaiian waters is very high: over fifty thousand individuals.

SHORT-BEAKED COMMON DOLPHINS *(Delphinus delphis)*

Short-beaked common dolphins have not yet been documented in Hawaiian waters, but I suspect that at some point in the near future they will be. They are an open-ocean species, found in a broad swath of the North Pacific not too far north of the archipelago, and they also use tropical waters to the east and southeast in the eastern tropical Pacific, as far as Central America. If not approached closely, they are likely to be confused with one of the species of *Stenella,* as they have a relatively long rostrum and a three-toned color pattern. Their rostrum is relatively short and robust, similar to a striped dolphin, and the most obvious distinguishing feature is a tan-colored patch on the side extending from the eye to below the dorsal fin. Their dorsal fin is often bicolored as well, with lighter-colored frosting in the middle of the fin. I am hoping that the first time they are confirmed in Hawaiian waters will be during one of our surveys, but I suspect they are more likely to be documented by observers on longline vessels, particularly during the winter when cooler, more productive waters move farther south toward the islands.

›››

A short-beaked common dolphin photographed off California in October 2008. This species has never been documented in Hawaiian waters but is likely to be at some point in the future, most likely in more productive waters to the north of the islands. Photo by Jim Cotton/SWFSC.

THE BEAKED WHALES

The first beaked whale I ever saw in Hawaiian waters was just minutes after leaving Honokōhau Harbor on the Kona Coast of Hawai'i Island, in April 2002. Just 2 kilometers from shore, a lone whale surfaced just a few hundred meters away from the boat. Amazingly, she approached the boat and we were able to get photos. As she circled the boat, we decided to try to get a skin sample for genetics, and using a crossbow and a biopsy dart we were successful. Without any teeth visible to aid us in identification, we did not know for sure what species of beaked whale this was, but genetic analysis of the skin sample confirmed that it was a

A group of five Blainville's beaked whales off Kona, May 22, 2009. Photo by Deron S. Verbeck/iamaquatic.com.

female Blainville's beaked whale. From the photos we could see that she had once been bitten by a large shark and had many scars from smaller shark bites. We have used these scars to identify her many times over the years, with the most recent sighting in 2014, all off Hawai'i Island.

Almost a quarter of the ninety species of cetaceans recognized worldwide are beaked whales, members of the family Ziphiidae. Beaked whales are distributed in deeper waters of the world's oceans. They are a primitive group of whales, having split off from the other toothed whales somewhere between 16 and 23 million years ago. They range in size from the pygmy beaked whale, reaching a maximum length of about 3.8 m (12′5″), to the Baird's beaked whale, reaching a maximum length of about 11.7 m (~38′). As their name implies, they have a distinct beak or rostrum, similar to many of the oceanic dolphins. With one exception, they have a greatly reduced number of teeth. One species found only in the Southern Ocean has a full set in both upper and lower jaws, two species have just two pairs of teeth, and the remainder have only a single pair of teeth located in the lower jaw. In all of those with just a single pair of teeth, they erupt only above the gums in adult males.

As a group, they are poorly known for several reasons: they typically live in the open ocean, usually far from shore, with some species remaining submerged for an hour or more at a time, and many are fairly cryptic in their surfacing patterns, making them difficult to detect even when at the surface. There are several species that have been extensively studied in multiple areas of the world, including Cuvier's and Blainville's beaked whales in Hawai'i, primarily because they inhabit the slopes of oceanic islands that rise steeply from the seabed. Although there are twenty-two species of beaked whales currently recognized, several new species have been described in the last thirty years, and it is likely that more will be described in the next thirty years.

Of the twenty-two currently recognized species (possibly twenty-three by the time you read this), fourteen are members of the genus *Mesoplodon*. These species are particularly difficult to tell apart unless photos are obtained of the head of an adult male that show the shape of the jaw and placement of the teeth, which are visible for some species even when the mouth is closed. Otherwise a skeletal specimen or genetic sample needs to be obtained to confirm species.

There is one other way to identify the species. Recent acoustic research indicates that beaked whale echolocation clicks differ from other species of whales or dolphins (in that they are frequency-modulated, upswept pulses), and each species (at least those so far recorded) produces distinct species-specific clicks. Unfortunately, echolocation clicks have not been positively recorded from all of the ten recognized species of beaked whales in the North Pacific. Hawaiian waters are known to be home to three species of beaked whales: the Cuvier's beaked whale, Longman's beaked whale, and one species of *Mesoplodon,* the Blainville's beaked whale. Two different types of beaked whalelike echolocation clicks have been recorded from Hawaiian waters that may represent other species of *Mesoplodon* that have not yet been visually documented, so it is likely that one or more additional species of beaked whales will be discovered in Hawaiian waters in the near future. Based on the acoustics and the known distributions of different species, the most likely candidates are the ginkgo-toothed beaked whale and the Hubbs' beaked whale.

We've had 141 sightings of beaked whales in Hawaiian waters through January 2016, 77 of which were Cuvier's beaked whales and 53 of which were Blainville's beaked whales. We've had just a single sighting of Longman's beaked whales, also known as the tropical bottlenose whale, and ten more sightings of beaked whales that we could not confirm to species.

CUVIER'S BEAKED WHALES *(Ziphius cavirostris)*

Cuvier's beaked whales have the widest distribution of any species of beaked whale in the world, being found from cold Antarctic waters through the tropics and into deep subarctic waters. The first Cuvier's beaked whale documented in Hawaiian waters was from a skeleton found on a beach at Ka Lae, the southernmost point of Hawai'i Island, in 1950.

For a beaked whale, Cuvier's can be quite easy to spot. They occasionally leap completely clear of the water, and given their size—up to 7 m (~23′) long—such breaches can be seen for miles. Breaches have occurred in nine out of our seventy-seven encounters with this species in Hawai'i, and in many cases a distant breach is what first alerted us to the group. Because of their wide distribution and tendency to be found on continental and island slopes, Cuvier's beaked whales are one of the best-known species of beaked whales in the world. Much of the research that has been undertaken has been driven by trying to understand why Cuvier's beaked whales appear to be so sensitive to impacts from naval sonar. Our research on this species in Hawaiian waters includes the longest-term photo-identification study of Cuvier's anywhere in the world, as well as studies of diving behavior, habitat use, and movements, all recorded using tags. Although we don't see them very often, the population of Cuvier's beaked whales off the island of Hawai'i may be the best-known population of this species anywhere in the world, given the length of the study and the diversity of methods we are using to study them.

Identifying Features and Similar Species

At birth, Cuvier's beaked whales are about 2.7 m (8′10″) long, and the longest reliable measurement of an adult was about 7 m (~23′). When young they appear a fairly uniform gray or brown in color, except for a darker spot around the eyes; while the skin is gray, they are often completely or partially covered by a film of brown diatoms, a type of algae. As they age, Cuvier's beaked whales in tropical waters become covered in more and more white spots, almost appearing polka-dotted. These spots are actually white scars, typically oval but sometimes rectangular in shape, that are healed bites from cookie-cutter sharks, a species that also inhabits deep waters in tropical areas like Hawai'i. Based on long-term resightings from our work, we know some of these scars are visible for at least nineteen years, and given that they spend a large proportion of their time deep in the water column where cookie-cutter sharks live, Cuvier's become progressively more and more covered with white scars as they age.

Distinguishing adult males from females can be accomplished in a couple of different ways. Like most other beaked whales, the teeth of Cuvier's beaked

A pair of Cuvier's beaked whales breaching off Kona, October 21, 2013. Even from this distance, it is often possible to identify individual beaked whales, using the large number of spots on the side from cookie-cutter shark bites. The female on the right is HIZc007. Photo by Annie B. Douglas.

whales erupt only through the gums for adult males. They have just two teeth, located at the tip of the lower jaw, and the lower jaw extends slightly beyond the upper jaw, so these teeth once erupted are continuously exposed and visible if the male's head is above the water's surface. Sometimes these teeth act as anchoring sites for stalked barnacles, so the white or cream-colored teeth may be partially or completely covered with reddish-brown barnacles. The males use these teeth for fighting with other males, and thus they also acquire extensive linear scars on the head and front half of the body. Even without viewing the teeth, if an individual has extensive linear scaring on the body it can probably be considered an adult male. The skin on the head and front half of the body of older adults of both sexes becomes progressively whiter as they age; this extensive white coloration is usually much more obvious in adult males.

Cuvier's beaked whales, like other species, often expose their beak upon surfacing, and a good view of the head is enough to confirm species because of their short rostrum and upturned mouthline. The head is most often exposed just before a long dive, when Cuvier's often lunge out of the water, exposing much of the body. Views of the head typically require getting close; telling beaked whale species apart when the head is not seen is a bit more difficult. In Hawai'i, the most difficult identity challenge is discriminating between Cuvier's and Blainville's beaked whales, but there are several somewhat subtle characteristics that can be used to tell them apart when the head is not seen or when an individual is lacking the extensive white coloration of an adult male. There is typically more arch or curve to the back of a Cuvier's beaked whale when surfacing, while Blainville's appear to break the surface (and dive) at a more shallow angle, without arching the back. When viewed from in front (or behind), the back of a Cuvier's is quite broad and relatively domed, while Blainville's are less robust and more laterally compressed, having a somewhat peaked back when viewed from in front or behind. On a normal surfacing, when the dorsal fin of a Cuvier's is clear of the water, the head and blowhole are usually submerged, while for Blainville's beaked whales both the dorsal fin and blowhole are usually clear of the water at the same time. Proportionately, the dorsal fin of a Cuvier's beaked whale appears smaller in comparison to the amount of back visible, and this can also be used to discriminate them from Blainville's. If the beak isn't seen, Cuvier's and Longman's beaked whales could be confused in the field, although given the rarity of

Longman's beaked whale groups around the main Hawaiian Islands, this is unlikely to happen often. The much larger group sizes of Longman's (typically thirty or more individuals) in comparison to Cuvier's (typically one to four individuals) is probably the fastest way of telling them apart.

Cuvier's beaked whale adult male (top), adult female (middle), and calf (bottom). The extensive linear scarring and erupted teeth can be used to discriminate adult males from adult females. Adult males tend to have more extensive white coloration than females; however, older adult females can also show similar coloration. The oval white scars are from cookie-cutter shark bites and accumulate with age. Illustrations by Uko Gorter.

A mother and calf pair of Cuvier's beaked whales off Kona, August 20, 2012. The female, HIZc018 in our catalog, was first documented off the island in September 2002. Photo © Dan J. McSweeney/Wild Whale Research Foundation.

An adult male Cuvier's beaked whale lunging at the surface prior to a long dive off Kona, April 15, 2006. One tooth is visible at the tip of the lower jaw, surrounded by purple-colored stalked barnacles. Photo © Dan J. McSweeney/Wild Whale Research Foundation.

Habitat Use, Movements, and Abundance

Cuvier's beaked whales have been sighted extensively among the Northwestern Hawaiian Islands, in offshore waters, and around the main Hawaiian Islands. Around the main Hawaiian Islands, however, all of our sightings of this species have been off Hawai'i Island, and there are only a few sightings (mainly from aerial surveys) off other islands. Although they have been sighted in depths as shallow as 780 m, sighting rates for this species off Hawai'i Island are highest in depths from about 1,500 to 3,500 m.

It is usually possible to tell Cuvier's and Blainville's beaked whales apart even from directly behind the animals. Cuvier's have a broad, rounded back, while Blainville's beaked whales (inset) have a laterally compressed body that appears peaked. Both of these photos were taken off Kona. Cuvier's beaked whale photo by Melissa Evans-Shontofski. Blainville's beaked whale photo by author.

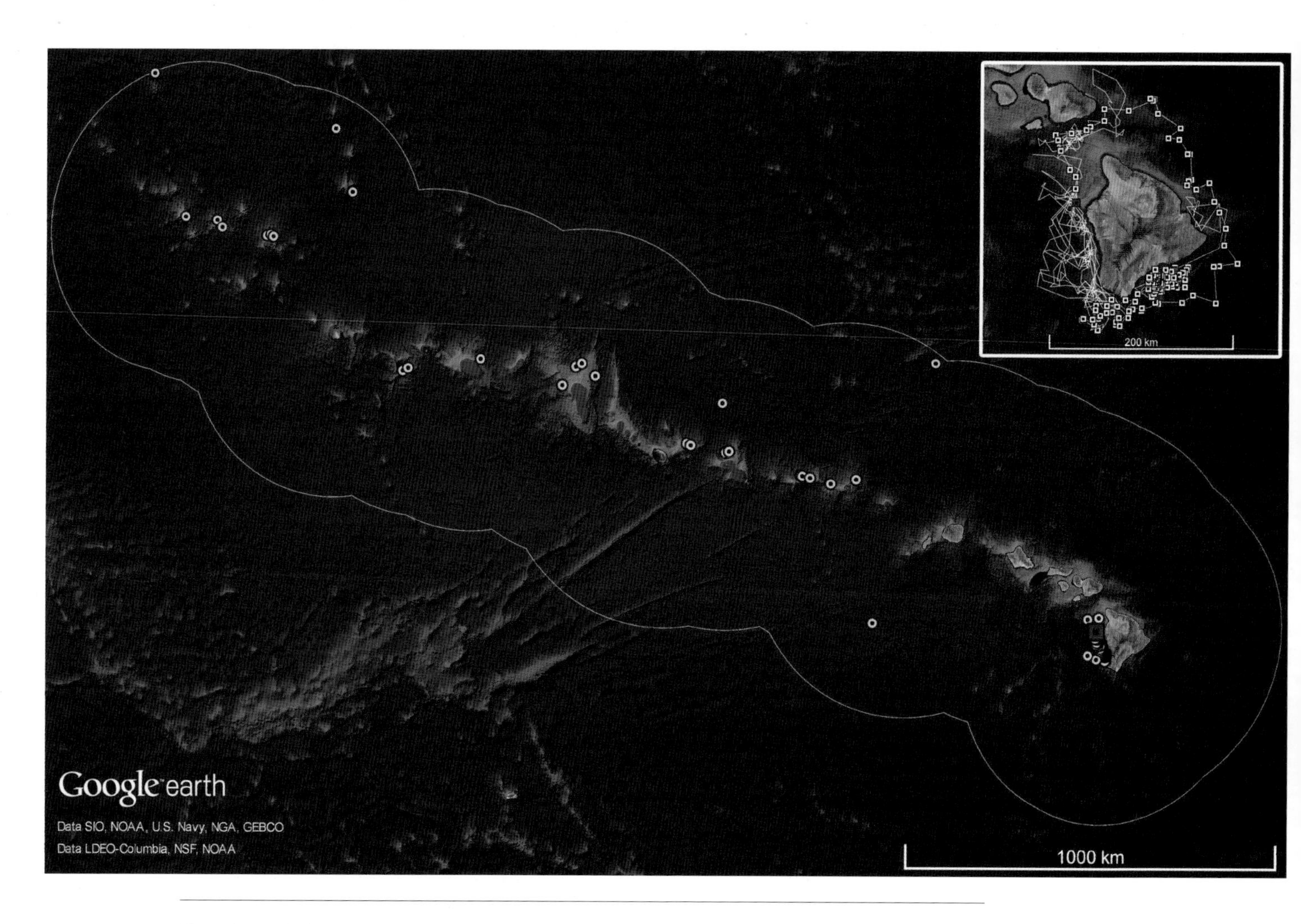

Cuvier's beaked whale sighting records (yellow circles) in Hawaiian waters. The inset shows tracks from ten satellite tagged Cuvier's beaked whales, all tagged off the west side of Hawai'i Island. The points (white squares) shown represent the locations of one tagged adult female over a thirty-three-day period in October and November 2011, with the tagging location shown in red.

Although our photo-identification catalog of Cuvier's beaked whales contains only about a hundred individuals, high resighting rates and long-term resightings (from ten to twenty years for seven different individuals and multiple years for another nineteen individuals) indicate there is a resident population off the island of Hawai'i. Based on resightings of photo-identified individuals, adult females seem to show a higher degree of fidelity to the island than adult males: they are resighted more often and over longer periods (up to twenty years so far) than adult males (up to ten years so far). From the photo-identification data, we've been able to estimate that only fifty-two individual Cuvier's beaked whales used the west side of Hawai'i Island over a four-year period, and covering all of Hawaiian waters there is an abundance estimate of 725 individuals from the 2010 NMFS survey.

We have movement data from ten individuals satellite tagged off Kona in six different years, including two adult males and seven adult females, and all have remained generally associated with Hawai'i Island, the south sides of Kaho'olawe and Maui, or the east side of Maui. The three individuals with tag attachments that lasted the longest—from thirty-two to forty-four days—moved the greatest distances, with one individual circumnavigating the island, one moving to the south and east of Maui, and one spending considerable time off the east side of the island east and south of Hilo. With additional tag deployments for longer periods, we are likely to document a larger range for this population than the data currently show. Whether such resident populations occur elsewhere in Hawaiian waters is unknown, but during the 2010 NMFS survey, Cuvier's were more concentrated around the slopes of the Northwestern Hawaiian Islands than in open-ocean waters. This suggests that there might be one or more island-associated populations in the Northwestern Hawaiian Islands as well. Although a lot of our surveys off Kaua'i, Ni'ihau, and O'ahu have been in relatively shallow waters where Cuvier's beaked whales are unlikely to be found, we've had enough effort in deepwater areas off those islands that we should have had sightings of Cuvier's. The fact that we have not suggests there are no island-associated populations in the western main Hawaiian Islands. Why this appears to be the case is unknown, but this species is particularly sensitive to impacts from naval sonar, and most of the U.S. Navy training activity in Hawai'i occurs off the western main Hawaiian Islands.

Predators and Prey

Like almost every species of cetacean in Hawaiian waters, large sharks, including both tiger sharks and white sharks, are likely predators of Cuvier's beaked whales, at least on occasion, and there is one adult individual in our catalog that has a healed wound just behind the head from a large shark bite. Killer whales are also likely to be occasional predators, although this has not been documented in Hawaiian waters.

Cuvier's beaked whales do most of their foraging at depth and rarely bring prey to the surface where they can be detected, but over the years we have recovered several squid from the water around Cuvier's beaked whales. It's likely that the squid were caught at depth and then brought up and either lost or discarded. Knowledge of the diet of this species comes primarily from stranded animals. Cuvier's feed primarily on deepwater squid, typically larger species of squid than consumed by Blainville's beaked whales. Cuvier's, like other beaked whales, are thought to be suction feeders: they have two throat grooves that allow them to stretch and expand their throat, sucking prey into their mouths. One adult female in our catalog is missing most of her upper jaw and can't close her mouth. Despite this, she seemed robust and had a calf with her last time we saw her, so such a deformity does not appear to have impacted her feeding or her mating.

Life History and Behavior

Little is known about the life history of any species of beaked whales except two that have been hunted extensively: Baird's beaked whales and northern bottlenose whales. For Cuvier's beaked whales, we don't know how old they are when they give birth, how often they have calves, or how long they live. We've seen only one newborn Cuvier's beaked whale in Hawai'i, in July 2008, and just four other individuals that were small calves but not newborn when first seen. One adult female first seen off Kona in 2004 (HIZc007 in our photo-identification catalog) has been documented with two different calves, one in 2004 and one in 2011, and in both cases the calves stayed with that mother for at least two years. HIZc007 was seen in 2008, 2009, and 2010, all without calves present, so the interval between having her two calves was over six years. Whether that is typical for Cuvier's beaked whales is unknown.

Probably the most unusual aspect of the behavior of Cuvier's beaked whales is their extremely long and deep dives: they hold the record for dive duration and dive depth of any mammal, with dives recorded off California to 2,992 m and 137.5 minutes. In Hawaiian waters, dives of up to 2,800 m and 88 minutes have been documented from our satellite-tagged individuals. It is not just the extremes that are amazing but that they make deep dives (over 1,000 m and 60 minutes in duration) day and night, nonstop, and likely year-round. Research has

An adult female Cuvier's beaked whale that is missing her rostrum, off Kona, April 23, 2009. Although this type of defect would appear to incapacitate an animal that feeds using suction, by her robust appearance and an accompanying calf (in May 2012), the missing rostrum does not appear to affect either her feeding or reproduction. Photo by author.

shown that foraging occurs during these deep dives, with the whales using echolocation to find and capture prey in the dark waters at depth. Cuvier's beaked whales do most of their echolocation when they are foraging below about 500 m depth; it is thought that they are remaining silent near the surface in part to minimize alerting potential predators, such as killer whales, of their presence.

Cuvier's beaked whales are not particularly social. Their average group size of 2.1 individuals is among the smallest of any species of toothed whale in Hawaiian waters; only pygmy sperm whales (1.5 individuals) are recorded in smaller average group sizes. Of our seventy-seven sightings, the most frequently encountered size is one. Lone individuals represent more than a third of all sightings. The largest group we've encountered in Hawaiian waters was of five individuals. Despite the small group sizes and infrequent sightings, a lot is known about Cuvier's social organization—who travels with whom, for how long, and why. One reason is that it is easy to determine the sex and relative age of individuals from photographs, and, while interactions among individuals may not be observed very frequently, their outcomes (in terms of the frequent linear scars on adult males) are obvious for years. Analyses of associations indicate that Cuvier's beaked whales have no long-term companions. The longest association observed was between a mother and calf and was only a couple of years, likely reflecting the period the calf was dependent. Larger groups of four or five individuals usually contained two adult males and two or three females and juveniles. In groups with more than one adult male, the males have remained separated and often have been observed tail lobbing (slapping their tails against the water's surface). The separation of the males, combined with all the tooth rakes, suggests they are competing over access to females, but sperm competition—where more than one male may mate with a receptive female and the males' sperm cells battle it out—may also be important for this species.

Interactions with other species of whales or dolphins are uncommon. The only species we've documented Cuvier's in close association with were short-finned pilot whales; on two occasions we've seen Cuvier's that appeared to be closely following them. We've seen them nearby other species, including one sighting near Blainville's beaked whales where the two groups were only a few hundred meters apart, but even though these two species were in the same area there was no indication they were interacting in any way.

A Cuvier's beaked whale diving off Kona, May 24, 2009. Photo by Deron S. Verbeck/iamaquatic.com.

Conservation

There is evidence of multiple populations of Cuvier's beaked whales in Hawaiian waters: an open-ocean population, a small resident population off the island of Hawai'i, and likely also an island-associated population in the Northwestern Hawaiian Islands. Despite this, only a single stock is recognized for all of Hawaiian waters, and this in itself may put the small Hawai'i Island population at risk. The Cuvier's beaked whale is thought to be the most sensitive to impacts from high-intensity sonars of any species of whale or dolphin. In a number of areas around the world, atypical mass strandings of Cuvier's beaked whales, often also involving other species of whales or dolphins, have occurred in association with naval sonar use. These strandings have occurred off Greece, in the Canary Islands off the coast of Africa, off Guam, and in the Bahamas. Such strandings might occur simply when an animal is trying to get away from a high-intensity sound, effectively blundering into shallow water and beaching. However, for Cuvier's, their extreme diving behavior likely puts them at particular risk, effectively making them more vulnerable to decompression-like symptoms if they dramatically alter their diving or surfacing behavior in response to a perceived threat such as sonar. In cases where there are small resident populations, such stranding events could clearly impact local populations, and the Hawai'i Island population is one of those that may be at risk from impacts, given the occurrence of U.S. Navy sonar testing and training in Hawaiian waters. Tagging studies have shown that Cuvier's react to very low levels of anthropogenic noise, typically by rapidly leaving the area. Yet in a naval training area off southern California, Cuvier's beaked whales do continue to use the area despite repeated exposure to sonar, although whether their foraging or other behaviors are impacted is not known. That area has a relatively high abundance of Cuvier's and is likely an important foraging area, making it a location they are unlikely to permanently avoid, even with regular use of naval sonar. This also suggests that individuals that are repeatedly exposed to sonar may moderate their reactions and potentially be at a reduced risk of injury or death. Conversely, naïve individuals, such as those off islands where the U.S. Navy rarely trains such as Hawai'i Island, may be at greater risk. A recent study has shown a decline in the Cuvier's beaked whale population off the west coast of the United States, although the cause or causes of the decline are not known. This does suggest that concern is warranted, particularly because a decline in the abundance of a species so rarely encountered will be difficult to detect, to say the least.

BLAINVILLE'S BEAKED WHALES *(Mesoplodon densirostris)*

Blainville's beaked whales have the second-widest distribution of any species of beaked whale, ranging from warm temperate waters and through the tropics, in both the northern and southern hemispheres. They were first identified in Hawai'i in 1961, when two individuals stranded on Midway, along with a Cuvier's beaked whale, over a three-day period. The first good underwater photos of this species in the wild were taken off the Kona Coast of the island of Hawai'i in the late 1970s and early 1980s, before any researchers were studying them in the wild. They are found both in the open ocean and on the slopes of oceanic islands. It is on these island slopes that this species has been studied extensively: off the Bahamas since 1991, in Hawai'i since 2002, and off the Canary Islands since 2003. As a result of these studies, Blainville's are one of the best-known species of beaked whales in the world. Although our work with beaked whales in Hawai'i began in 2002, our photo-identification catalog for this species spans almost thirty years, with photos provided by Dan McSweeney of the Wild Whale Research Foundation, working off Hawai'i Island since the late 1980s.

Identifying Features and Similar Species

Blainville's beaked whales are about 2 to 2.5 m (6′6″ to 8′3″) long at birth and may reach up to 4.7 m (15′5″) in length. Like other beaked whales, the dorsal fin is relatively small and located about two-thirds of the way back on the body. At birth Blainville's are largely light gray in color, with darker pigmentation around the eye, although they quickly acquire a variable coating of brown diatoms. As they age, Blainville's beaked whales in tropical areas become spotted from cookie-cutter shark-bite scars, just like other beaked whales, and the bites heal white and remain visible for many years. Based on a long-term resighting of a photo-identified individual, we know the scars can remain white for at least twenty-one years. Even though the scars fade slowly with time, the whales acquire them much more frequently than they lose them and become more and more polka-dotted as they age. Cookie-cutter bites can often go right through the dorsal fin, leaving the individual with a permanent hole through the fin.

The heads of Blainville's beaked whales often clear the surface when they come up for a breath. This juvenile has black spots around the mouth that are likely scars from hooks on squid tentacles. This photo was taken off Kona, October 21, 2013. Photo by Amy Van Cise.

The sex of adult Blainville's beaked whales can be determined based on the scarring on the body: while both males and females have oval scars from cookie-cutter shark bites, only adult males, such as this individual, have extensive linear scars from fighting with other males. This photo was taken off Kona, November 22, 2006. Photo by author.

Blainville's are sexually dimorphic in a couple of ways. Adult females are slightly larger than adult males, although this isn't obvious in the field. The posterior half of the lower jaw in both males and females is highly arched upward (a trait that can be used to discriminate Blainville's from other species in the genus *Mesoplodon*), but in males it is taken to an extreme: the arch of the lower jaw can extend above the upper jaw, so that the highest points on the head are actually the arched lower jaws and the teeth that erupt from the apex. The males use these teeth for fighting, often gouging tissue away from the base of the teeth of other adult males, so the amount of each tooth that is visible is quite variable. One

Blainville's beaked whale adult male (top), adult female (middle), and calf (bottom). The lower jaw of the adult male is highly arched and the two teeth erupt from the highest point of the jaw. Adult males also accumulate linear scars from fighting with other males. Illustrations by Uko Gorter.

An adult male Blainville's beaked whale as the head breaks the surface. The two erupted teeth are obscured by stalked barnacles, which anchor onto the exposed surface of the teeth. This photo was taken off Kona, July 10, 2008. Photo by Annie B. Douglas.

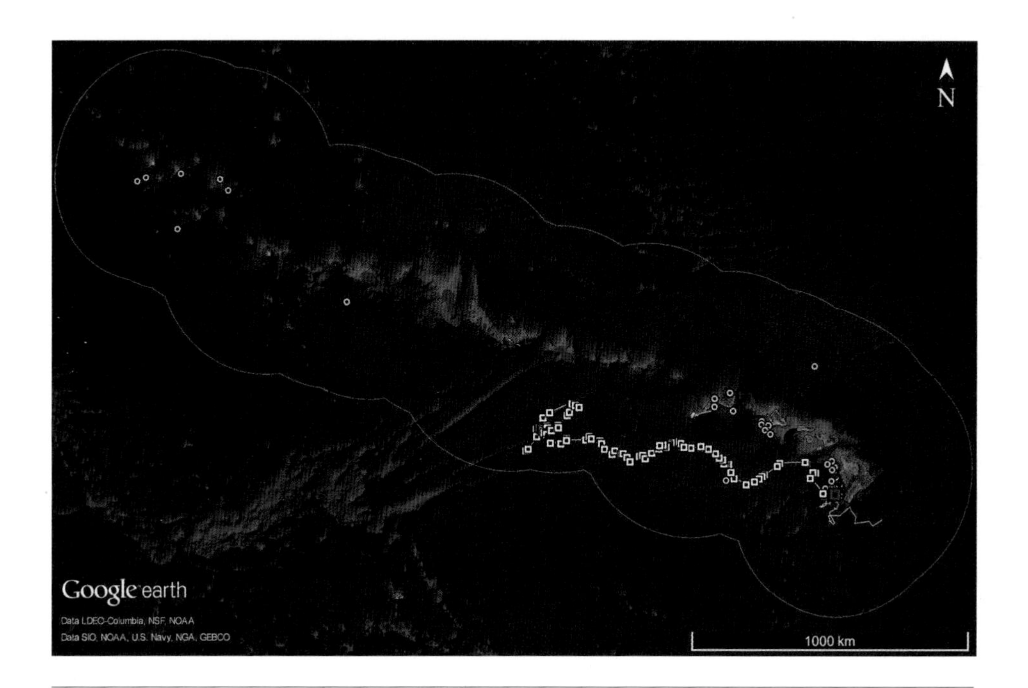

Blainville's beaked whale sighting records (yellow circles) and tracks from three satellite tagged individuals, all thought to be from the pelagic population (lines). Locations for a forty-day track of an adult male from the pelagic population are shown with white squares, with the tagging location shown in red.

adult male in our catalog has only a single erupted tooth visible, and the arched jaw that is missing the tooth is lower than the other, suggesting it may be missing the tooth entirely or the top was broken off at a young age. Particularly if tissue is missing from around the tooth, they act as anchoring sites for stalked barnacles, so they often appear to have a purple or reddish-brown growth on the teeth. Because of this extensive fighting, even without viewing the head, adult males can be distinguished from adult females by the extensive linear scarring on the back.

In Hawai'i, Blainville's beaked whales are most likely to be confused with Cuvier's beaked whales, unless the head is visible: the beak or rostrum of a Blainville's is longer than a Cuvier's, and the arched mouthline is distinct. Even without a view of the head, the two species can be distinguished based on the relative size of the dorsal fin and amount of back showing upon surfacing, how much the animal arches when it surfaces, and, if viewed from behind, whether the back is domed or peaked (see the section on Cuvier's beaked whales for details). They could also be confused with ginkgo-toothed beaked whales or Hubbs' beaked whales, although neither of these has been positively documented in Hawaiian waters. Adult male Hubbs' beaked whales have a distinct white cap on the head, which should be diagnostic, but discriminating females or either sex of ginkgo-toothed beaked whales would probably require a genetic sample. In the case of ginkgo-toothed beaked whales, the external morphology and coloration of adult males is poorly known. Blainville's beaked whales could also be confused with small baleen whales such as minke whales, although they tend to arch their back much higher than Blainville's do and are rarely seen around the main Hawaiian Islands.

Habitat Use, Movements, and Abundance

Although Blainville's beaked whales have been sighted in offshore waters and around the Northwestern Hawaiian Islands, most of the *Mesoplodon* sightings in those areas were from the NMFS large-vessel surveys, and the animals were not observed closely enough to confirm species. Blainville's have been recorded among the main Hawaiian Islands much more frequently, with sightings off Ni'ihau, Kaua'i, O'ahu, and Hawai'i Island. While I know of no confirmed sightings off Maui Nui (the four-island region of Maui, Moloka'i, Lāna'i, and Kaho'olawe), two individuals tagged off Hawai'i Island did move north and to the west, off the north side of Maui, one of them moving as far as the north side of Moloka'i, so they do use those areas.

Long-term resightings of individuals, habitat use, and satellite tag data all indicate that there is a resident, island-associated population of Blainville's beaked whales off the island of Hawai'i. Our photo-identification catalog of Blainville's beaked whales is relatively small (one hundred distinctive individuals documented since 1991), and there are long gaps between sightings. Yet forty of the one hundred individuals have been seen more than once, and seven different individuals have been resighted off the island over spans of more than ten years, with one resighted over a span of twenty-one years. Almost all the individuals

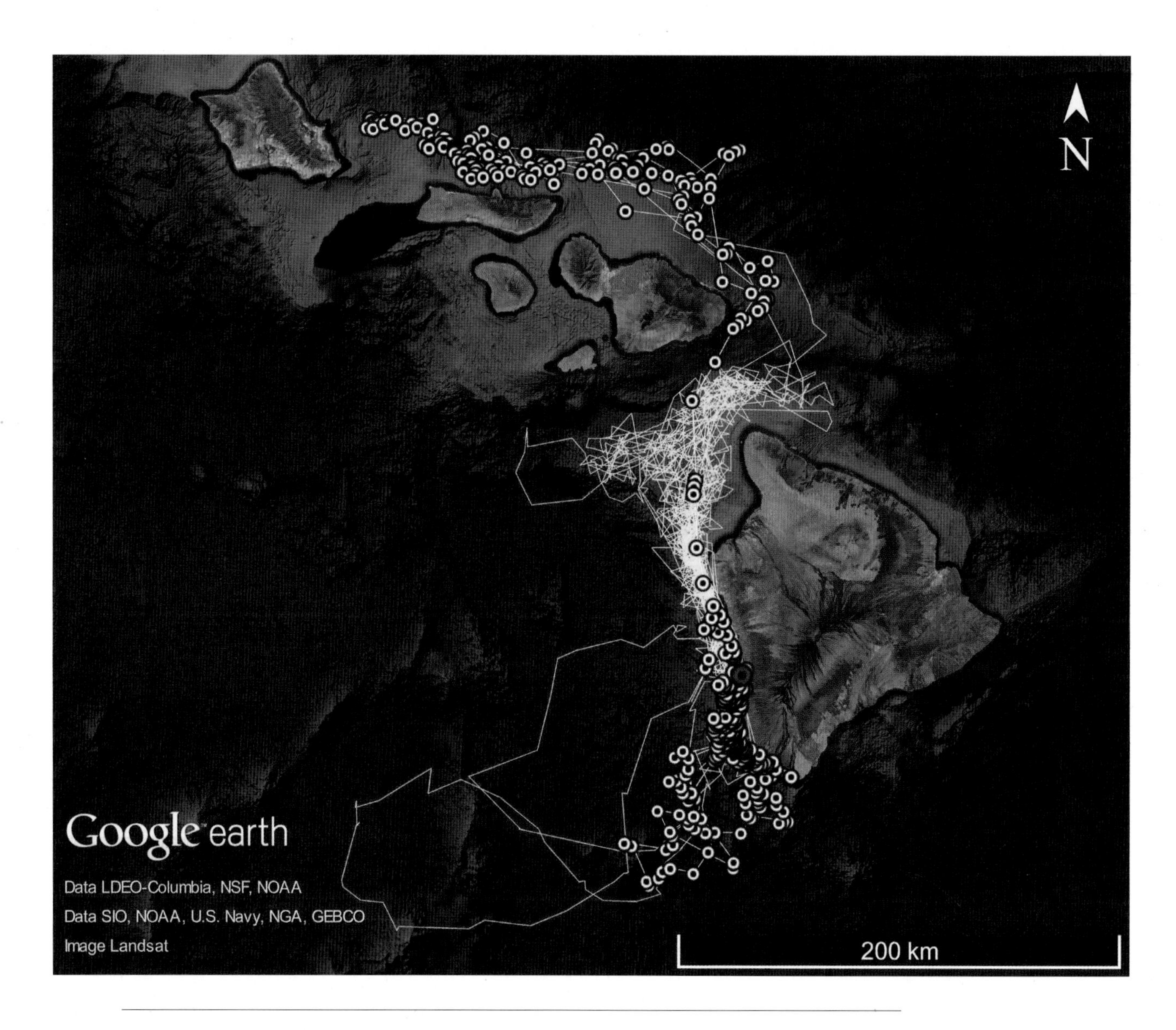

Tracks (yellow lines) of thirteen satellite tagged Blainville's beaked whales from the Hawai'i Island resident population. Locations for one adult male over a 170-day span are shown with white circles, with the tagging location indicated by a red circle.

photo identified off Hawai'i Island, and all those seen more than once, are part of the same social network—not seen together all at once but linked by association with others in the social network. We have satellite tag data from fifteen individuals off the island, thirteen of which are part of the resident social network and two that aren't. One that wasn't in the social network was tagged offshore in water more than 3,800 m deep and tracked over forty days. During that time, the individual moved over 1,000 km from the island to the west, generally meandering over areas about 5,000 m deep—the abyssal plain. Comparing the movements of the individual to eddies that spin off Hawai'i Island, it appeared that the whale did pay some attention to those large-scale oceanographic features in its meandering. This individual is clearly part of an open-ocean population and has not been resighted off the island. The thirteen satellite tagged individuals that are all part of the resident social network remained generally associated with the islands, with the exception of two individuals that made offshore excursions, at different times, after which they both returned to the island slope. While Blainville's beaked whales have been sighted on both the leeward and windward sides of the islands, most of the time the thirteen tagged individuals remained off the west (leeward) coast of Hawai'i Island.

The only individuals tagged off Hawai'i Island that have left the island slope and moved into waters near another island were both adult males. From our photo-identification work we knew that adult males show less fidelity to the island, and our tagging work supports that as well. We've tagged two Blainville's in the channel between Kaua'i and Ni'ihau, but sadly one of the two tags did not function. An eight-day track we obtained from the other individual showed it moving to the south of Ni'ihau to the area around Ka'ula Island, to the southwest. Although our catalog of Blainville's photo identified off Kaua'i and Ni'ihau is small, none of the individuals has been resighted, suggesting that there may not be a resident population off those islands as there is off Hawai'i Island. We've seen Blainville's in water as shallow as 360 m as well as in very deep water, but sighting rates for this species peak between about 500 and 1,000 m depth, before dropping off in deeper waters. Sighting rates climb again in depths over 3,500 m, likely representing sightings from the open-ocean population.

Abundance of Blainville's beaked whales for all of Hawaiian waters was estimated from the 2010 NMFS survey, at just over 2,100 individuals, although the estimate was not very precise. Off the island of Hawai'i, the estimated number of individuals that used our study area over a four-year period was 140 individuals, although this included both individuals from the resident, island-associated population and the offshore population.

Predators and Prey

Although attacks by killer whales have not been observed in Hawaiian waters, one photo-identified Blainville's beaked whale in our catalog has tooth-rake scars on the dorsal fin that appear to be from a killer whale attack. Three different individuals photographed off Hawai'i Island have bite marks that appear to be from large sharks, probably either tiger or white sharks. Like Cuvier's beaked whales, Blainville's do most or all of their foraging deep in the water column and thus observations of prey are rare, limited to a few occasions when squid have been recovered in the vicinity of this species. Although their primary prey are thought to be deepwater squid, from reports elsewhere Blainville's are also thought to feed to some extent on deepwater fish. Blainville's also suction feed, and we have one healthy-looking adult female in our catalog with a deformed upper jaw, another indication that such deformities don't seem to prevent beaked whales from feeding.

Life History and Behavior

Like Cuvier's beaked whales, not a lot is known about the life history of this species, as they were never taken in whaling operations, and only a small number of stranded animals have been aged. However, information is beginning to come in from the long-term photo-identification studies being undertaken in several areas. Males and females are both thought to be sexually mature at about nine or ten years of age, and females give birth to their first calf between ten and fifteen years of age. The interbirth interval seems to be about three years, and calves are weaned between two and three years of age.

Blainville's, like Cuvier's beaked whales, are known to forage deep in the water column. In Hawai'i, Blainville's have been documented diving to 1,520 m, with dive durations as long as 83.4 minutes, and dives over 1,000 m and an hour in

We've used suction-cup-attached tags to obtain high-resolution, short-term information on diving behavior or acoustics. This tag on a juvenile Blainville's beaked whale tagged off Kona, July 10, 2008, recorded information on dive depths. Photo by author.

A mother and calf pair of Blainville's beaked whales surfacing off Kona, July 10, 2008. Photo by Daniel L. Webster.

duration are not uncommon. Blainville's do such deep foraging dives during both the day and at night, indicating that whatever they are feeding on, they seem to remain at depth at all times. Since Blainville's live on island slopes with the highest sighting rates between 500 and 1,000 m, many of these deep dives are likely reaching the bottom. Blainville's beaked whales have to prepare for these long and deep dives. In the time they are at the surface prior to long dives, we've recorded them taking between thirty-eight and forty-one breaths over a period of four or five minutes. They also seem to have to recover at the surface after long dives: we've recorded them taking thirty-two breaths at the surface after a long dive. Between shorter and shallower dives, they may surface to breathe only three to ten times.

Blainville's beaked whales are a bit more social than Cuvier's; the average group size from our sightings is 3.7 individuals. Of our fifty-three sightings, the most frequently encountered group was of three individuals, and lone individuals were documented on only six occasions (about 12 percent of sightings). The largest group we've seen among the islands was eleven individuals. But like Cuvier's beaked whales, Blainville's don't seem to have any strong or long-lasting associations. Blainville's are not particularly social with other species of whales or dolphins either. We've seen them near four other species—short-finned pilot whales, rough-toothed dolphins, bottlenose dolphins, and Cuvier's beaked whales—but they have never paid any obvious attention to the other species.

Conservation

Like Cuvier's beaked whales, the biggest conservation concern for Blainville's beaked whales is the impact of high-intensity naval sonars. Blainville's have been involved in multispecies stranding events in response to naval sonars in the Bahamas and the Canary Islands, and such stranding events can have population-level impacts, particularly if there are small resident populations, such as in Hawai'i. Despite evidence for both a small island-associated resident population off Hawai'i Island and an offshore population, only a single stock of Blainville's beaked whales is currently recognized for Hawaiian waters. This in itself puts the Hawai'i Island population at risk. Potential impacts of naval sonar are calculated over all of Hawaiian waters, assuming there is a large population with individuals moving freely. In reality at least some individuals are resident to a relatively small area, where concentrated sonar use could potentially impact most or all of the population. Research with Blainville's beaked whales on a naval range in the Bahamas has shown that when sonar is used, whales will typically stop echolocating (and therefore stop foraging) and move away from the sonar source, sometimes taking days before they move back to the area where they had been foraging. It is likely that such cessation of foraging and increased movements can have population-level impacts, even when strandings are not observed or documented. A study by Diane Claridge in the Bahamas has shown that Blainville's beaked whales in an area with frequent exposure to naval sonar may have fewer calves; thus the population may decline or have less potential to increase. Similar reductions in echolocation and movements off the Kaua'i U.S. Navy range have been documented, but whether there have been population-level effects is unknown.

LONGMAN'S BEAKED WHALES *(Indopacetus pacificus)*

In the past there has been reference to southern bottlenose whales (*Hyperoodon planifrons*) or a "tropical bottlenose whale" in the central tropical Pacific, including Hawai'i, but in recent years it has become apparent that that these were actually Longman's beaked whales, a species that was first described in 1926 and that was effectively rediscovered in recent years. A sighting of "bottlenose whales" was reported from off the island of Hawai'i in the 1970s, and another group of forty or fifty large but unidentified beaked whales was seen in the same area in November 2000. The first confirmed record of Longman's beaked whales in

An adult male Longman's beaked whale (right) with one erupted tooth visible, in Hawaiian waters, July 22, 2015. Photo by Jessica K. Taylor, courtesy of the International Whaling Commission. Taken during the IWC POWER (Pacific Ocean Whale and Ecosystem Research) cruise; see www.iwc.int/power.

A Longman's beaked whale in Hawaiian waters, north of the Northwestern Hawaiian Islands, September 11, 2010. Photo by Sophie Webb/SWFSC.

Hawaiian waters was a sighting during the 2002 NMFS survey in offshore waters to the southwest of the Northwestern Hawaiian Islands. In our work, we've had a number of reports of sizable groups of large beaked whales off the island of Hawai'i, but we have seen the species only on a single occasion—a group of about thirty-five individuals in August 2007. One stranding of a Longman's beaked whale has been recorded in Hawai'i, a juvenile that stranded in March 2010 on Maui. In the 2010 NMFS survey there were four more sightings of this species, all in offshore Hawaiian waters. The most recent sighting of Longman's beaked whales in Hawaiian waters was a group of about 110 individuals seen on a joint International Whaling Commission/Japanese research cruise in July 2015, in offshore waters south of Nihoa.

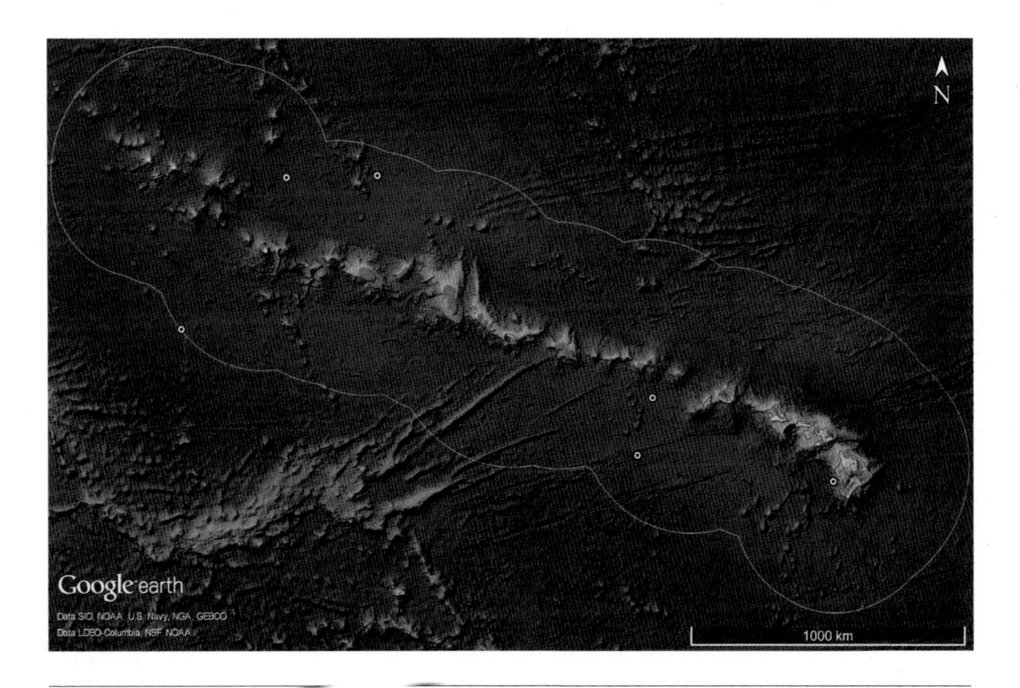

Confirmed Longman's beaked whale sighting records in Hawaiian waters. There are several more reports of groups of beaked whales that are likely this species off Kona, but no photographs were taken to confirm the species.

A fast-moving group of Longman's beaked whales in Hawaiian waters, August 31, 2010. Longman's often move much faster than other species of beaked whales in Hawaiian waters, throwing up a characteristic spray of water. Photo by Sophie Webb/SWFSC.

Compared to either Cuvier's or Blainville's beaked whales, Longman's beaked whales have a very bulbous forehead, and a distinct, sharply demarcated beak, and their dorsal fin is taller relative to body size. Longman's are the only one of the three that usually has a distinct blow visible when they exhale. Relatively few specimens of Longman's beaked whales have been examined, so even basic information such as maximum length is unknown. From sightings at sea it is clear that Longman's are relatively large beaked whales, perhaps reaching lengths of 9 m (29′6″), but the largest specimen measured to date was 6.5 m (21′4″). Like the other Hawai'i species, adult Longman's beaked whales tend to be heavily scarred with white oval scars from cookie-cutter shark bites, and these scars likely remain visible for years, so this species could easily be photo identified if enough photographs were obtained. Coloration is generally gray or light brown. Adult males have two teeth at the tip of the lower jaw, although it is difficult to see the teeth when adult males expose their head at sea.

Sightings suggest that they tend to move faster and more directionally than either of the other Hawai'i beaked whale species and that they travel in much larger and more cohesive groups. Group sizes in Hawaiian waters have ranged from 18 to 110 individuals. When traveling very fast, they often throw up a characteristic rooster-tail of spray. Despite the rarity of sightings, Longman's may be more abundant in Hawaiian waters overall than any other species of beaked whale. In the 2010 NMFS survey, when the observers were officially "on-effort," there were three sightings of Longman's with large average group sizes (~sixty individuals), compared to a single sighting of seven Blainville's beaked whales and two sightings of lone Cuvier's beaked whales. The resulting abundance estimate for Longman's beaked whales for all of Hawaiian waters, based on some complicated math and the amount of area covered, was over 7,600 individuals.

GINKGO-TOOTHED BEAKED WHALES *(Mesoplodon ginkgodens)* AND HUBBS' BEAKED WHALES *(Mesoplodon carlhubbsi)*

Neither of these species has been confirmed in Hawaiian waters, but evidence from analysis of acoustic recordings made at four different locations in Hawai'i suggests that one or both species may occasionally occur here. For those species of beaked whales for which recordings have been made, each produces species-specific echolocation clicks. The clicks of most species of beaked whales are well above the hearing range of humans, but recording systems have been developed to document these clicks at sea. Autonomous recording systems called High-frequency Acoustic Recording Packages (HARPs) have been deployed off Kona, Kaua'i, Cross Seamount (to the southwest of Hawai'i Island), and at Pearl and Hermes Atoll (in the northwestern Hawaiian Islands). Recordings of the distinctive echolocation clicks of Blainville's and Cuvier's beaked whales have been made at some of these locations, but two other types of clicks, both thought to be from beaked whales, have also been recorded. Based on recordings of such clicks elsewhere in the North Pacific, including off the west coast of the United States and Mexico and in the western and central Pacific, one of these two clicks is thought to come from ginkgo-toothed beaked whales, and the other is thought to come from Hubbs' beaked whales.

Ginkgo-toothed beaked whales are found in the tropical waters of both the western and eastern Pacific, typically to the south of Hawaiian waters. The clicks that are thought to possibly come from ginkgo-toothed beaked whales have been recorded extensively at Cross Seamount and also occasionally off Kona, Kaua'i, and Pearl and Hermes Atoll. Hubbs' beaked whales are found both off Japan and off the west coast of North America, typically inhabiting temperate waters to the north of Hawai'i. The clicks that are thought to come from Hubbs' beaked whales have been recorded once in the Northwestern Hawaiian Islands, at Pearl and Hermes Atoll. Confirming these species in Hawaiian waters will likely require good head photos of Hubbs' beaked whales and a genetic sample of ginkgo-toothed beaked whales. Adult male Hubbs' beaked whales are quite distinctive, with a bulge on the melon that is typically pigmented white and a highly arched jaw (although not as high as in Blainville's beaked whales) with large teeth that erupt from the apex of the arch. The coloration and external appearance of adult male ginkgo-toothed beaked whales is poorly known, however. Every time we see a group of small beaked whales in Hawai'i, we hope they may be one of these two species, but so far, no luck.

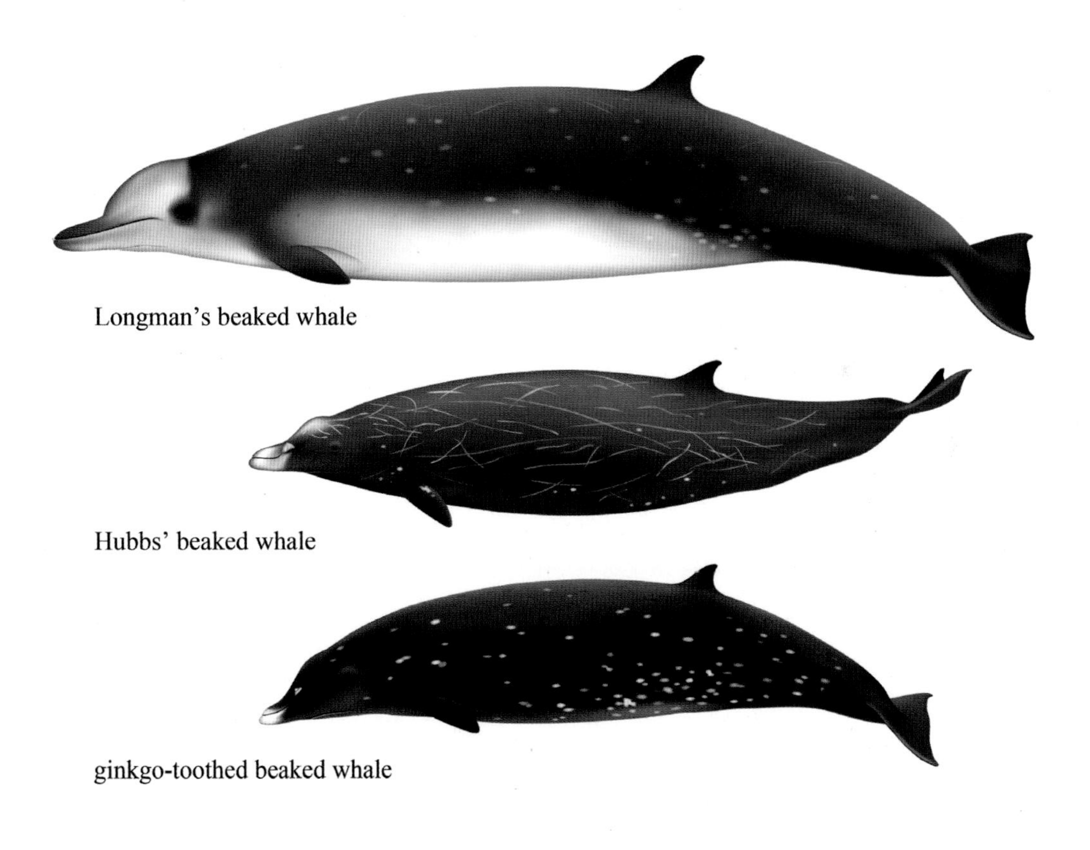

While neither Hubbs' beaked whales (middle) nor ginkgo-toothed beaked whales (bottom) have been visually documented in Hawaiian waters, based on species-specific echolocation clicks and the known distributions of these species it is likely that both are at least occasionally found in Hawaiian waters. A Longman's beaked whale (top) is shown for comparison. All three illustrations represent adult males. Illustration by Uko Gorter.

THE SPERM WHALES

About 30 million years ago, just a few million years after the evolutionary split between baleen whales and toothed whales, one group diverged from the rest of the odontocetes: the sperm whales. At one point in their evolutionary line, there were many different species of sperm whales, but only three (or possibly four) survive today in what is considered the "superfamily" of sperm whales. These include the largest of all toothed whales, the iconic sperm whale from the pages of *Moby Dick* (the only member of the family Physeteridae), and two of the smaller species, the dwarf and pygmy sperm whales, members of the family Kogiidae. All three of these species are found in Hawaiian waters. There is some evidence that there are

A friendly sperm whale off Kona, December 1, 2014. Photo by Deron S. Verbeck/ iamaquatic.com.

actually two species of dwarf sperm whales, one in the Atlantic Ocean and another throughout the Pacific and Indian Oceans, but only one is currently recognized.

There are a number of similarities between sperm whales and their smaller cousins, the dwarf and pygmy sperm whales. All three have blunt heads and relatively small, underslung jaws. The lower jaws have numerous teeth, but the number of teeth in the upper jaw is greatly reduced, or in some cases teeth are absent altogether. Unlike other species of whales or dolphins, the blowhole is not centrally located on the head. In the dwarf and pygmy sperm whales, the blowhole is slightly off center to the left, while in the sperm whale it is located at the front left corner of the head. A behavior that is common to all three species is logging, which involves floating motionless at the surface, and they can be seen in this resting behavior more than any other species of whale or dolphin in Hawaiian waters.

Dwarf and pygmy sperm whales are closely related (members of the same genus, *Kogia*) and were not recognized as different species until 1966; earlier references referred to both species as pygmy sperm whales. Although it is almost impossible to see in the field, they both have a band of light pigmentation on the side of the head that is often referred to as a "false gill," and it has been suggested that this makes them appear similar to a shark, perhaps as a way of reducing the likelihood of an attack from sharks. When startled, and occasionally at other times, both dwarf and pygmy sperm whales are known to release a cloud of "ink," although it is actually a slurry of fecal material, stored in an enlarged section of the colon, where it can be retained for up to ten days. This enigmatic behavior was witnessed in 1985 by an observer on a tuna purse seine vessel in the eastern tropical Pacific. The observer watched a mother and calf dwarf sperm whale that were accidentally encircled in the purse seine net, along with a school of spotted and spinner dolphins and the associated tuna. Whenever a dolphin would approach the mother and calf, the mother would release a cloud of reddish material into the water, and then both mother and calf would hide within the cloud until it dissipated.

These species are all typically found in deep waters, with the smallest, the dwarf sperm whale, the most likely to be seen around the main Hawaiian Islands. In our study off the islands through January 2016, we'd had 137 sightings of these three species, 86 of which were of dwarf sperm whales, 45 of sperm whales, and only 6 of pygmy sperm whales. This is despite the fact that, owing to their size, sperm whales are much easier to spot than dwarf sperm whales. Adult female

Dwarf (top) and pygmy (bottom) sperm whales. The two species can be distinguished based on relative dorsal fin size and position, as well as head size. Illustrations by Uko Gorter.

sperm whales are about 9–12 m (29′5″–39′4″) in length compared to a maximum of 2.7 m (8′10") for a dwarf sperm whale, and they are usually sighted at greater than a kilometer away, compared to just a few hundred meters for dwarf sperm whales. Interestingly, of the three species, pygmy sperm whales end up stranded on shore the most often—about four times as often as dwarf sperm whales and about 20 percent more often than sperm whales in Hawai'i—despite being so rarely seen at sea. All strandings of these species in Hawai'i have been of single animals, with the exception of single mother-and-calf pairs of both dwarf and pygmy sperm whales. Sperm whales have been extensively studied in the wild in many areas of the world, while dwarf and pygmy sperm whales are rarely studied; most of what is known of these two species comes from stranded animals. The exception, for dwarf sperm whales, is in Hawaiian waters: the Kona Coast of Hawai'i Island is the site of the only long-term photo-identification study of this species in the world.

DWARF SPERM WHALES *(Kogia sima)*

Even in areas where dwarf sperm whales are known to live, boaters rarely see them, or if they do, they often don't know what they are seeing, as dwarf sperm whales spend little time at the surface and usually offer only a quick glimpse before disappearing from sight. In Hawai'i they were first reported as stranded individuals, one on O'ahu in 1966 and one on Lāna'i in 1987. It wasn't until 2002 that there were a number of sightings that could be confirmed to species in Hawaiian waters. At sea, dwarf sperm whales are most likely to be seen floating motionless at the water's surface (logging) in waters deeper than 500 m, often

An adult female dwarf sperm whale off Kona, May 17, 2012. This individual is HIKs020 in our photo-identification catalog, a female first seen off the island in November 2004 and most recently documented in October 2014. Photo by Daniel L. Webster.

An adult male dwarf sperm whale that live stranded on Kaua'i, August 27, 2009. The photo shows the small underslung jaw and the light band of pigmentation between the eye and the flipper that has been described as a false gill. After restranding multiple times, this individual was euthanized. Photo by Kim Steutermann Rogers.

alone or in small groups (two to four individuals). They float with part of the back and the dorsal fin visible. Unless startled, they usually remain at the surface for several minutes before disappearing and are often not seen again. Most of the time, they will sink out of sight rather than rolling at the surface as do most species of dolphins, although occasionally they perform a slow roll at the surface. They rarely approach boats and are difficult to follow, as their surfacing patterns are difficult to predict. While they may be pointed in one direction when they are logging at the surface, once they have submerged the direction they travel may have little to do with the direction they were originally pointed. When logging, they often orient either directly up or down the prevailing swell; presumably this is a more comfortable way to log, rather than rolling side-to-side in a swell. When there are more than two together, they are often loosely associated, with up to several hundred meters between pairs of individuals. Any sort of aerial behavior (leaping, breaching, spyhopping) is extremely rare; in our eighty-six sightings, we've seen animals leap out of the water only three times. They are usually seen only in very calm sea conditions; in rough seas (anything with whitecaps), they are almost impossible to spot.

Identifying Features and Similar Species

At birth, dwarf sperm whales are about a meter (3′3″) in length, and the largest adult documented was about 2.7 meters (8′10″). The heaviest dwarf sperm whale I'm aware of weighed 259 kg (571 lbs). At sea they appear dark gray, although they do have a lighter underside. As with other species of cetaceans, such countershading is probably an adaptation to reduce predation by sharks or killer whales hunting deeper in the water column. Their dorsal fin is located at about the midpoint of the back, although when logging at the surface the back half of their body is submerged and the dorsal fin appears farther back. The sexes differ slightly, with adult males being a bit larger than adult females, but this is not something that can be seen in the field.

There are four other similar-sized species that may log at the surface showing only the back and dorsal fin: the closely related pygmy sperm whale, pygmy killer whales, melon-headed whales, and less often rough-toothed dolphins. Observations of both dwarf and pygmy sperm whales tend to be very brief, making discriminating between these two species most difficult. Many books suggest that it is almost impossible to tell the two species apart, but there are some subtle differences between them that can be recognized from photos or from a good view before they dive. When logging, the back of a dwarf sperm whale is relatively flat (level) from the head to the dorsal fin, while the back of a pygmy sperm whale has a distinct bulge between the dorsal fin and head. In part this bulge reflects the proportionately much larger head size of a pygmy sperm whale. If the blowhole and the front of the head can be seen (either by eye or in a photo), the relative head size can also be used to discriminate between the two species.

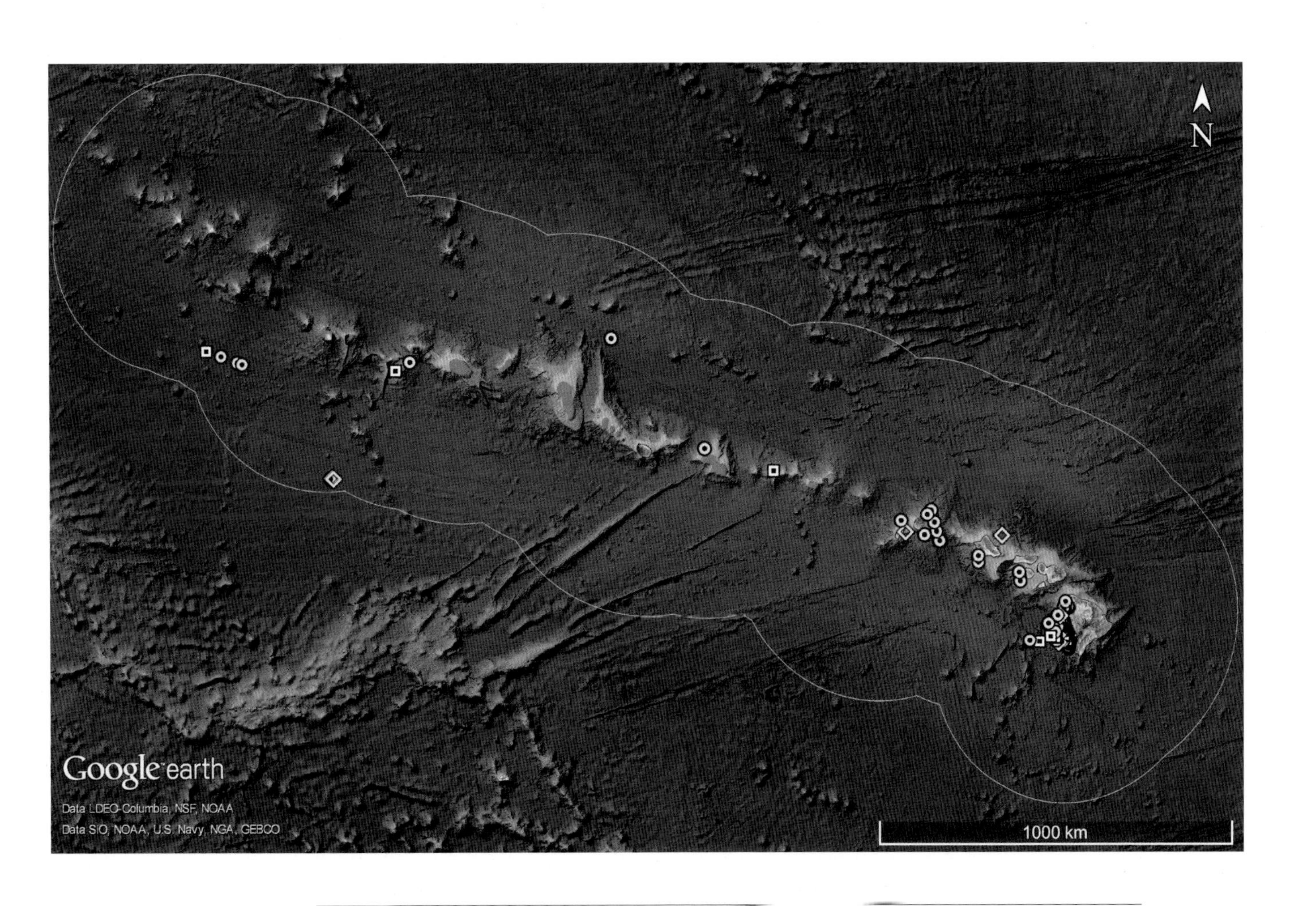

Sighting records of dwarf sperm whales (yellow circles), pygmy sperm whales (yellow diamonds), and cases when the exact species couldn't be determined (white squares) in Hawaiian waters.

The relative size and shape of the dorsal fin can also be used to tell the two species apart: the dorsal fin of a pygmy sperm whale is shorter (less than 5 percent of the total length) and located farther back on the body, while the dorsal fin of a dwarf sperm whale is taller (greater than 5 percent of the total length) and thus larger in proportion to the amount of back typically visible. The tip of the dorsal fin of the dwarf sperm whale also tends to be pointed compared to rounded in the pygmy sperm whale, although older animals tend to have a lot of damage to their dorsal fin, so fin shape can vary dramatically.

Of the other three species of small odontocetes that log at the surface, pygmy killer whales and melon-headed whales also have rounded heads, so if only seen while logging they might be a subject of mistaken identity. Unlike dwarf sperm whales, however, both pygmy killer whales and melon-headed whales are typically found in larger groups (5–20 for pygmy killer whales, 50–350 for melon-headed whales), and they do not tend to dive for long periods. Rough-toothed dolphins do log at the surface occasionally, but they often approach boats, and if their long beak is seen they can easily be identified. Short-finned pilot whales also log at the surface, and juveniles overlap in size with dwarf sperm whales, but juvenile pilot whales are unlikely to be seen alone or in small groups without adults present that could be easily distinguished.

Habitat Use, Movements, and Abundance

Dwarf sperm whales have been documented in offshore waters, off the Northwestern Hawaiian Islands, and offshore of most of the main Hawaiian Islands, although they appear to be less common off the western main Hawaiian Islands—Ni‘ihau, Kaua‘i, and O‘ahu—than off the eastern islands. The highest sighting rates of dwarf sperm whales around the main Hawaiian Islands are in areas with depths between 500 and 1,000 meters, but they have been documented in depths as shallow as 106 m and as deep as 4,700 m.

Almost nothing is known about movements of dwarf sperm whales. Although we have been trying to get close enough to this species to deploy a satellite tag since 2013, they seem to have a large personal space, and getting inside it is extremely difficult. Trying to approach closely enough to remotely deploy satellite tags requires staying with groups for extended periods, and it is only possible to do so if the group isn't disturbed. In more than half of our encounters with this species, we've lost the group within fifteen minutes, but we are learning how to stay with groups without disturbing them. During these longer encounters, lasting up to an hour and thirty-three minutes for our longest, groups tend to move slowly (1–2 km/hour) along a meandering path. Now we often lose a group because the weather deteriorates, not because they are avoiding us, and we are typically able to get good photos of most or all of the individuals we encounter.

Almost all adult-sized individuals have distinctive dorsal fins, and individuals can be followed over time with photographs. Although we have a photo-identification catalog of this species with over a hundred individuals, most were photographed off the island of Hawai'i; thus we can't yet assess movements among islands using photo identification. From resightings of individuals off

Three dwarf sperm whales logging at the surface off Kona, April 10, 2010. Photo by author.

Hawai'i Island, we have found one that moved as far as 58 kilometers along the coast between sightings.

Dwarf sperm whales are probably one of the more abundant species of cetaceans in Hawaiian waters. In the 2002 NMFS survey covering all of Hawaiian waters, their estimated abundance was over seventeen thousand individuals, the highest abundance estimate of any species. Despite their abundance, they are identified only infrequently for a variety of reasons. With their small size and inconspicuous behavior, they are difficult to spot in anything other than ideal water conditions, and they also take long dives, so it is easy to miss them. In our surveys around all the main Hawaiian Islands they are the sixth most frequently encountered species of odontocete overall and the fifth most frequently encountered species in waters less than 1,000 m deep, representing almost 8 percent of all sightings. Relative abundance varies among the islands, however, and they appear to be more common off Kona than elsewhere. Based on photo identification of distinctive individuals off the west coast of Hawai'i Island, there appears to be a resident slope-dwelling population, with high resighting rates. Fourteen different individuals have been documented in multiple years, with several seen over spans of five or six years. One very distinctive adult female, HIKs020 in our photo-identification catalog, has been documented in eight different years over a ten-year span, from 2004 through 2014.

Predators and Prey

Based on scars, dwarf sperm whales are probably frequently attacked by large sharks, but unlike many of the smaller species of odontocetes they often survive. Many of the bite wounds are on the dorsal surface around the dorsal fin, which suggests that, upon detecting a looming shark, they may roll over to protect their more vulnerable belly, and given how well healed the wounds are, this strategy seems to work. Although not recorded in Hawaiian waters, predation by killer whales on dwarf sperm whales has been documented elsewhere. They are frequently bitten by cookie-cutter sharks and often have both fresh and well-healed wounds visible when good photographs are obtained, although the scars repigment to the same color as the background and are difficult to see. These bites are probably nothing more than a nuisance to the whales, seldom causing significant

Healed scars on the head of a dwarf sperm whale indicate that this individual survived an attack by a large shark. The relatively fresh oval wounds are from cookie-cutter shark bites. This photo was taken off Kona, April 10, 2010. Photo by author.

damage, although we have seen some with bites completely through the dorsal fin, healing and leaving a hole.

Dive depths have never been measured for dwarf sperm whales, and little is known about diet in Hawai‘i, although based on examination of stomach contents of stranded animals elsewhere, they consume a diverse range of small squid and occasionally mesopelagic fish, as well as shrimp.

Life History and Behavior

Most of what is known about the life history of dwarf sperm whales comes from a study of stranded animals in South Africa. Males reach sexual maturity between 2.6 and 3 years of age, and maximum age documented for a male was 17 years. Females reach sexual maturity at 5 years of age and have been documented living to 22 years. Most interestingly, females are thought to give birth every year and can be simultaneously nursing a calf and pregnant with a new calf, a rare trait among toothed whales. Two newborn individuals have been documented in Hawai‘i, one in March and one in October, suggesting that if there is a breeding season it may not be strongly seasonal.

Dwarf sperm whales are not particularly sociable. Most individuals are observed alone, representing 30 percent of all of our sightings, and the largest group we've seen was only eight individuals. The average group size of dwarf sperm whales in Hawai‘i is a paltry 2.7 individuals. Only Cuvier's beaked whales and pygmy sperm whales seem to have more asocial tendencies—their average group sizes are only 2.1 and 1.5 individuals, respectively. That said, more than half the distinctive individuals we've documented off Hawai‘i Island have been linked together by association in the same social network, suggesting it is a relatively small community of individuals that probably all associate together, at least occasionally. Photo identification, however, reveals few repeated associations beyond weaning. One group of three individuals (one adult female, one subadult, and a second adult of unknown sex) was seen together two days apart. Two adults, one known adult female and one of unknown sex, were seen together almost exactly five years apart, although both individuals were seen alone or with others in the intervening years, suggesting that they did not spend that period together. Most likely their reassociation was just an infrequent event in a small population, rather than reflecting any particular bond between the two individuals. Dwarf sperm whales similarly seem to avoid (or at least not seek out) other species of whales or dolphins—we've never seen them interacting with other cetaceans, even when other species were in the vicinity.

Dwarf sperm whales seem to avoid public displays of affection; we witness very few social interactions among individuals. However, aerial observations in the Bahamas by researcher Diane Claridge revealed that, once below the surface, dwarf sperm whales may rapidly swim and chase each other like other odontocetes. Such spirited behavior is surprising, given their reserved nature at the surface.

Maximum dive durations are difficult to determine, as I suspect we often just lose individuals when they go on a really long dive, or we miss their subtle surfacing, particularly if they are facing directly away or toward us, when their profile above the water would be small. If we are eventually successful at tagging a dwarf sperm whale, we will learn a lot more about diving patterns in a very short period. In the two cases where newborn calves were documented, long dives of the mothers and neonates were only two to three minutes in duration. As the calf matures, both the mother and calf typically dive for periods of about seven minutes. Older animals dive longer, perhaps for fifteen minutes or more. Most of the time dwarf sperm whales are thought to avoid boats, but with patience they can be approached within about 50 or 100 m without apparently disturbing them. Once we had a calf that seemed to become curious about our boat, idling in neutral nearby; it started to approach, but the mother intervened and appeared to herd it away.

Conservation

Only a single population of dwarf sperm whales is recognized for all Hawaiian waters, although the long-term resightings of identified individuals solely off Kona, the relatively high resighting rates, and the concentration of sightings along the slope of the island all suggest there is a small resident population. Such resident populations may be particularly at risk from anthropogenic impacts. Dwarf sperm whales may be sensitive to high-intensity underwater sounds, and they readily exhibit negative reactions to large vessels. There have been several strandings that suggest the species may be impacted by high-intensity sonar. An unusual multispecies mass stranding in North Carolina in 2005 included thirty-three short-finned pilot whales, one minke whale, and two dwarf sperm whales, spread out over a broad area, and it occurred at the same time that the U.S. Navy was undertaking an exercise using MFA sonar offshore. In August 2009, a dwarf sperm whale live stranded and died on Kaua'i the same day a naval training event was being undertaken. In March 2014, there was a report of twenty-two dwarf sperm whales that died in association with dynamite fishing in the Philippines. I suspect one of the reasons dwarf sperm whales appear to be largely absent in the deeper water between Kaua'i and Ni'ihau is the concentration of naval sonar exercises that occurs there.

Fisheries interactions are likely impacting dwarf sperm whales to some extent as well—line injuries on the leading edge of the dorsal fin of some individuals photo identified off Hawai'i Island suggest they occasionally interact with fisheries. There have been three reports of dwarf or pygmy sperm whales being hooked on fishing gear in Hawaiian waters, including one pygmy sperm whale that was recently observed hooked in the longline fishery in offshore waters, although such hookings are infrequent.

PYGMY SPERM WHALES *(Kogia breviceps)*

Although distributed in deep waters worldwide, from temperate areas through the tropics, sightings of this species are rare, to say the least. They were first confirmed in Hawai'i when an individual was speared in Kahului Harbor on Maui in 1942; presumably the individual was in the harbor on its way to the beach to strand, as their normal habitat is offshore waters. In our work, pygmy sperm whales are one of the least-encountered species of odontocetes: through January 2016, we've seen them on only six occasions, representing far less than 1 percent of all odontocete sightings. Like dwarf sperm whales, pygmy sperm whales

Pygmy sperm whales have been the second most frequent species of toothed whale or dolphin to strand in Hawai'i. This individual, an adult male 3.07 m (10'1") long, stranded on Maui, April 25, 2007. Photo by Nicole Davis, NOAA Fisheries.

tend to log at the surface, but sometimes not even the dorsal fin is visible until they roll out of sight. With just the head and back visible and motionless at the surface, at a distance pygmy sperm whales are more likely to be taken for a small floating log than a species of small whale, until they roll and dive or sink out of sight.

As noted in the account for dwarf sperm whales, there are several ways to tell pygmy sperm whales from dwarf sperm whales. Although pygmy sperm whales are larger than dwarf sperm whales, reaching a maximum length of 3.8 m (12′6″) and a maximum known weight of 515 kg (1,135 lbs), judging size can be difficult in the field or from photographs. The dorsal fin of a pygmy sperm whale is smaller and located farther back on the body than on a dwarf sperm whale, and the dorsal fin tip is usually quite rounded. The relative head size of a pygmy sperm whale is much larger than for a dwarf sperm whale. Even when logging at the surface this can be obvious, as there is a slightly distinct "neck" behind the head, and the back between the head and dorsal fin has a distinct rounded arch, unlike the relatively flat back profile of a dwarf sperm whale.

We've seen pygmy sperm whales only once off Kaua'i, once off O'ahu, and four times off the island of Hawai'i. Three of those sightings were of lone individuals and the other three were pairs of individuals, suggesting this is not a very social species. Our sightings of this species tend to be very brief, with the individuals diving shortly after being spotted and not being seen again in four cases, and identifying them at a distance only through binoculars for another group. We've had just one encounter where we were able to stay with the group for a while, documenting dives that were fifteen minutes and twelve minutes long. During the 2002 NMFS survey, there was one sighting of a single pygmy sperm whale in offshore waters. There is an abundance estimate from the 2002 survey of over seven thousand individuals, but this species was not seen alive at all during the NMFS 2010 survey (one animal was seen floating dead), reflecting how difficult they can be to detect. Unlike most other species of odontocetes, there appear to be few sightings of this species by boaters or the general public, because individuals apparently dive before there is a chance to get a photo or a good enough look to confirm the species.

Despite the rarity of sightings, pygmy sperm whales are the second most frequently stranded species of odontocete in Hawaiian waters, representing more

Despite being quite abundant in Hawaiian waters, pygmy sperm whales are rarely seen alive, and this is actually a good view of one. It is very difficult to get close to this species, as they usually dive and are not seen again. This photo shows the distinctive bulge on the back between the fin and the head and the small, rounded dorsal fin set far back on the body. This photo was taken off Kona, May 16, 2012. Photo by Daniel L. Webster.

A pygmy sperm whale off Kona, May 16, 2012. This photo shows that the dorsal fin of a pygmy sperm whale is smaller in proportion to the amount of back visible than is seen on a dwarf sperm whale. Photo by author.

than 10 percent of all odontocete strandings. All but one of the strandings were lone individuals, and most were likely coming up on a beach due to sickness and subsequently died. Stranded individuals in Hawai'i have often had full stomachs, thus a fair amount of information is available on their diet here. Thirty-eight different species of squid have been documented from the stomachs of seven stranded individuals, and five of the seven stomachs also contained fish. Pelagic deepwater shrimp were also found in all of the stomachs and appeared to be an important part of the diet. Based on the habits of the prey species, it is likely that pygmy sperm whales are foraging between 600 and 1,200 m deep in the water column.

Like dwarf sperm whales, pygmy sperm whales are probably somewhat susceptible to impacts from high-intensity sonars. Two pygmy sperm whales stranded in the Canary Islands in 1988 as part of a multispecies atypical mass stranding event, associated with naval maneuvers. There were two strandings in Hawai'i in April 2007 that occurred following naval activity, although without knowing exactly where the animals were in relation to the activity, it is impossible to say for sure whether naval sonar caused the strandings.

SPERM WHALES *(Physeter macrocephalus)*

When sperm whales—*palaoa* in Hawaiian—washed onshore, their bodies were claimed by the chief. Their teeth were carved and worn as a pendant by the highest ranked *ali'i,* the chiefs or royalty. Sperm whales were also used as a source of food and oil for Native Hawaiians, and they were likely the most abundant whale in the deeper waters around the main Hawaiian Islands until the onset of industrial whaling. This species was one of the main targets of commercial whaling starting in the mid-eighteenth century, including in the waters of the central Pacific surrounding the Hawaiian Islands. In the twentieth century, an estimated 761,000 sperm whales were killed in industrial whaling worldwide. As a result, populations were reduced in most areas of the world, and the sperm whale was listed as endangered under the Endangered Species Act in 1970, one of only two species of odontocetes in Hawaiian waters that are listed as endangered.

The sperm whale is the largest toothed whale in the world, the only large whale likely to be seen throughout the year around the main Hawaiian Islands, and probably the most common large whale in deep offshore waters, even during the humpback whale breeding season.

Identifying Features and Similar Species

Sperm whales are about 4 m (~13′1″) long at birth, and as adults are strongly sexually dimorphic: adult females reach up to 12 m (39′4″) in length, while adult males can reach 18.3 m (60′). The head of a sperm whale is boxy and extremely large in proportion to its body, measuring about a quarter to a third of total body length. One of the most distinctive and unusual anatomical features is the asymmetrical location of the blowhole, located at the front of the head and on the left side, instead of being located in the middle of the head as is typical with most species. With calm winds and when viewed from a good angle, the bushy blow, up to 5 m (16′) in height for an adult male, projects forward and to the left; the sperm whale is the only species that has a blow like this. The dorsal fin of a sperm whale is also unlike any other species—it is a rounded hump about two-thirds of the way back on the body and is followed by a series of bumps on the dorsal surface of the tail stock.

A mother and calf sperm whale off Kona, October 31, 2013. This photo shows the rounded dorsal fin of the female and the blowhole of the calf, located at the front left part of the head. Photo by Daniel L. Webster.

Between their long, deep dives, sperm whales rest at the surface and replenish their breath. Usually when doing this they are moving at about 4 km/hour by slowly sculling their tail flukes, but sometimes they are nearly stationary. Despite this apparent lethargy, they occasionally leap completely clear of the water, with a large splash that can be seen for miles. One of the things we record for sightings is how far away we are when we first detect a group, and we've seen the splash of a sperm whale breaching at 8.4 km—over 5 miles. When initiating a long dive, sperm whales usually raise their flukes vertically above the surface and then slip below the air-water interface. They often defecate as they do this, leaving a large cloud behind in the water.

Given their size, sperm whales are likely to be confused primarily with the larger baleen whales such as humpbacks. The most obvious distinguishing feature from a distance is the spout: as noted, it projects forward and to the left

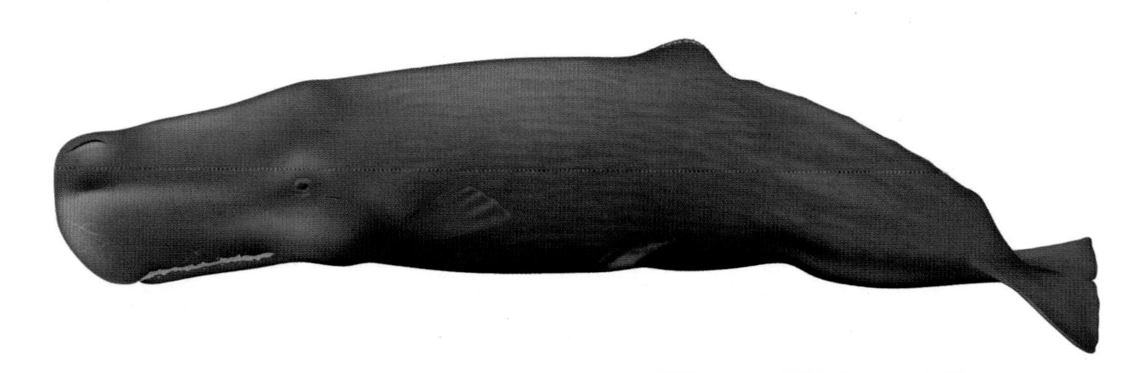

[Above] The sperm whale's head represents about one-third of its body. Illustration by Uko Gorter. [Opposite page, top] Sperm whales often fluke (raise their tails out of the water) just before a deep dive, and researchers use the shape of the trailing edge of the fluke to identify individuals. This whale was photographed off Kona, December 2, 2014. Photo by author. [Opposite page, bottom] A sperm whale breaching off Kona, May 22, 2013. The white pigmentation is actually scarring on the upper and lower jaws around the mouth. Photo by Brenda K. Rone.

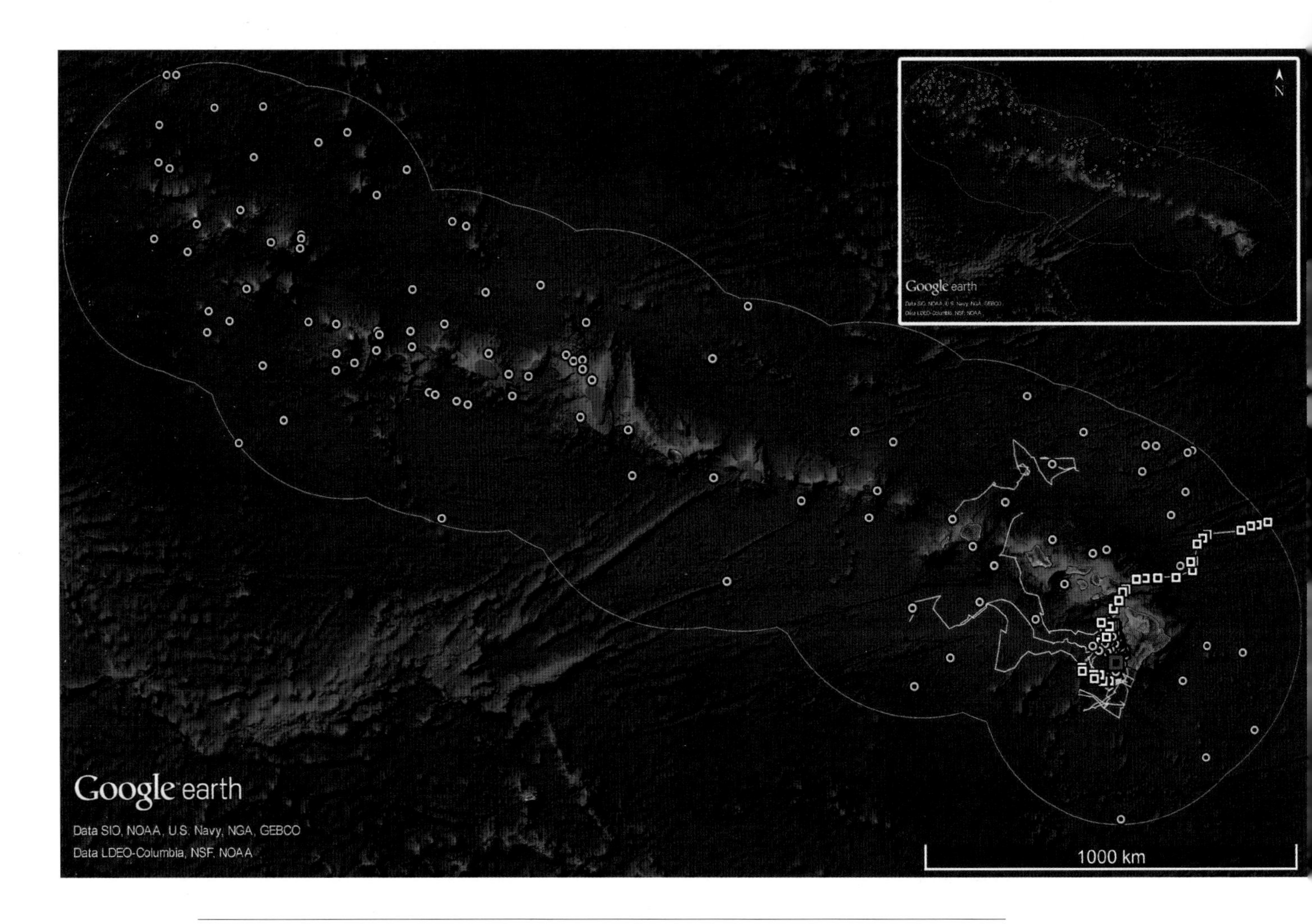

Sperm whale sightings (yellow circles) are distributed throughout Hawaiian waters. This map also shows tracks from twelve satellite tagged individuals (yellow lines); locations for one, a fourteen-day track of an adult female, are shown with white squares, with the tagging location shown in red. The inset shows the locations of 1,967 sperm whales killed in Hawaiian waters by Soviet and Japanese whalers between 1973 and 1976.

of the head. Their tendency to consistently remain visible at the surface is also a good clue—none of the baleen whales tend to do this. If the blow isn't easily visible, the boxcarlike head differs from the more pointed heads of most baleen whales and the rounded but fairly flat head of the humpback whale. The rounded dorsal fin of a sperm whale is also distinctive. Both humpback and sperm whales commonly raise their flukes in the air before diving. Obtaining a photo or a good view of the fluke shape and coloration will help to distinguish the two species, as humpbacks often have some portion of white on the undersides of the fluke, while sperm whale flukes are all dark. Humpback flukes have a distinct and often complicated trailing edge pattern, while sperm whales have less complicated and generally more rounded flukes.

Habitat Use, Movements, and Abundance

While sperm whales are found in deep waters from the Antarctic to the subarctic, their distribution and movement patterns differ between the sexes. Females and juveniles tend to remain in the tropics, subtropics, and warm temperate waters year-round, while males become nomadic, leaving the group they were born into by the time they are in their midteens. These dispersing subadult males first live in bachelor groups, but as they mature they move farther and farther from the tropics into more productive waters. As older adults, males spend most of their time at high latitudes and return to the tropics only occasionally for breeding. Sperm whales have been sighted throughout Hawaiian waters, including offshore areas, around the Northwestern Hawaiian Islands, and around the main Hawaiian Islands, and these tropical waters are primarily the home of groups of females and juveniles.

Around the main Hawaiian Islands, we've documented sperm whales in as shallow as 1,098 m, but they are seen at much higher rates in depths greater than 2,000 m. In depths greater than 4,000 m, they were the fourth most frequently encountered species of odontocete, representing more than 9 percent of all odontocete sightings. From the 2010 NMFS survey, their abundance was estimated at over 4,500 individuals for all Hawaiian waters.

We've been able to study the movements of sperm whales in Hawaiian waters using satellite tags. We've tagged twelve different individuals from nine different groups since 2009 and tracked their movements over periods ranging from six to fourteen days. One of the things that is obvious from the tag data is that sperm whales don't seem to be particularly associated with the islands but instead appear to move broadly. While we are tagging them relatively near shore (mostly off Kona but one off Kaua'i), only one of the groups stayed relatively close to the islands (over a nine-day period). There were certainly no consistent patterns in movements—tagged groups have meandered in all directions, with several traveling between the islands to the north and some going west or south. After being tagged, one adult female first moved west of Kona, then turned and headed to the east through the 'Alenuihāhā Channel and traveled to the northeast. By the time the tag stopped transmitting after fourteen days, the whale had moved almost 1,000 km and was in international waters over 500 km from where she was tagged.

Looking at all the tag data, recognizing that the tracks are relatively short, it appears that sperm whales around the main Hawaiian Islands are part of a broadly ranging population that likely extends across much of the central tropical Pacific. Despite the fact that they probably cover a lot of water, individuals do return to the islands occasionally. Jay Barlow at the Southwest Fisheries Science Center has a sperm whale fluke catalog that includes photos we've taken in Hawai'i, and we've started a catalog of distinctive dorsal fins, but one long-term resighting of a sperm whale off Kona came from two underwater photographers on whale-watching vessels. One animal with several distinctive white patches on the belly was photographed off Kona in October 2010 and again, by a different photographer, in February 2015. All the individuals we've tagged so far have been adult females or subadults, so we can't say anything about how long adult males remain in the area or how far they move. Adult males are seen relatively infrequently around the main Hawaiian Islands: in our forty-five sightings of sperm whales, we've seen only five adult males. Adult male sperm whales from Hawai'i likely migrate to forage in areas off Alaska and Russia, and genetic analyses throughout the central and eastern Pacific show that some sperm whales sampled off the Aleutians and in the Gulf of Alaska likely originated from the central Pacific around Hawai'i.

Predators and Prey

One would think that the largest toothed whale would be immune from attacks by killer whales, but even they are attacked on occasion, and at least one well-observed attack off the U.S. West Coast was successful. Unlike smaller cetaceans, sperm whales do not typically flee if attacked by killer whales. Instead, they form up in a very tight group and either face the attackers with jaws open or group into a rosette or marguerite (named after a flower) formation, heads together and tails out toward the attacking predators. Which strategy they use tends to depend on the social group, based on the frequency of killer whale tooth scarring on the tail flukes of sperm whale groups. I know of one case of killer whales around a group of sperm whales off the Kona Coast of Hawai'i Island, and although they weren't observed attacking, the sperm whales were bunched together in a very tight unit. As with most species of large whales, it is likely that sperm whale calves are at much greater risk from killer whale attack than adult whales. In theory, large sharks might be a risk as well, particularly to calves, but no attacks have been observed.

Several of the satellite tags we've deployed on sperm whales in Hawai'i have transmitted information on dive depths and durations, and from these we know that sperm whales are diving deep, and thus likely feeding, both during the day and at night. The rates at which they dive below 500 m are very similar between the day and night, as are maximum dive depths. The deepest we've documented sperm whales diving in Hawaiian waters is 1,552 m, deeper than any species other than Cuvier's beaked whales. Like Cuvier's, sperm whales mainly eat deep-water squid, and they catch an extremely wide range of species, varying in size from 100 grams to the giant squid, *Architeuthis,* with females that can weigh up to 900 kg (almost 2,000 lbs). Foraging dives of 1,000 m or more typically last from about thirty-five to sixty minutes in duration, and the longest dive we've recorded in Hawai'i lasted seventy-five minutes.

Life History and Behavior

Based on data obtained from whaling operations and long-term behavioral studies, the life history of sperm whales is well known relative to most other whales

A group of sperm whales logging at the surface off Kona, October 31, 2013. Seventeen individuals can be counted in this photo. Photo by author.

or dolphins. They tend to mass strand, which provides an additional wealth of information. Females first conceive at about nine years of age, gestation lasts about fifteen months, and females have one calf every four to six years thereafter. While males may be sexually mature in their teens, it is unlikely that they mate until their late twenties, when they first migrate back to warmer waters from their high-latitude foraging areas. Females probably live into their eighties or longer; longevity of males is not as well known, but they probably live at least into their sixties or seventies.

Although the average group size of sperm whales in Hawaiian waters is about six individuals, a quarter of our sightings were of lone individuals, and the largest group we've documented was of thirty-two whales, albeit spread out over a large area. Defining a group is difficult with sperm whales, however, as they can be extremely spread out over large distances when foraging. Seeing individuals separated by a couple of kilometers is not unusual, and one group of about fourteen individuals was dispersed over an area of about 15 by 4 km. In these cases, we typically see single individuals or mother/calf pairs, often diving for long periods, with each of these individuals or pairs separated by a kilometer or more. When at the surface, all tend to be headed in the same direction, and these dispersed groups often move through the area as a unit over time. The extremely loud echolocation clicks produced by sperm whales can be heard over tens of kilometers, and thus even when broadly dispersed, individuals are probably aware of the locations of others in these aggregations. Based on studies conducted elsewhere of individually identified sperm whales, these aggregations of individuals likely contain one or more long-term social units. Although males do disperse from their natal group and females occasionally switch groups, the units are thought to be generally stable over decades.

Most of the time, adult sperm whales either appear largely oblivious to boats or dive prematurely if a boat approaches too rapidly or too closely; juveniles, however, will often seem curious, approaching our boat and circling it. Photo-identification studies of this species tend to rely on photographs of the outline of the tail flukes, taken as the animal dives, so approaches for research purposes are generally done from directly behind an individual when they are breathing at the surface. In this way, they can be approached slowly from behind to try to get a photo, and as long as they aren't approached too closely they don't react. Sperm

Juvenile sperm whales often appear to be quite curious toward boats. This individual approached our research vessel off Kona and rolled just below the surface, with the eye open and visible, November 30, 2014. Photo by Brenda K. Rone.

whales' behavior toward boats is not always indifference or avoidance, however. There is the well-known case of the whaleship *Essex,* sunk by a sperm whale in the equatorial Pacific in 1820, which was the inspiration for Herman Melville's book *Moby Dick.*

Other ships have also been sunk by sperm whales, although usually these attacks have been provoked events involving whales that were being hunted, as with the *Essex.* Such is not always the case, however. On July 25, 2006, a 40-foot sailing yacht named the *Mureadritta XL* was sunk by a sperm whale several hundred miles north of Hawai'i. I spoke with the captain by phone shortly after the incident. He described spotting whales in the distance, but the yacht was on a course to give them a wide berth. One of the whales left the group and rammed the side of the yacht with its head, caving a large hole into the side. I was sent photos of where the whale had hit, and bits of gray skin were visible in the photo around the fractured fiberglass. The boat sank and all aboard were rescued, but this event reflects that whales are wild animals and their behavior is not always predictable or benign. Fortunately, in our work with sperm whales, even after tagging or biopsying individuals, we've never seen any evidence of aggressive behavior.

Conservation

Although sperm whales have been listed as endangered in the United States since 1970, commercial hunting of them by other nations continued in the North Pacific until 1987. In 1976, the United States declared a 200-nautical mile fishing exclusion zone, which went into effect March 1, 1977. Prior to that, hunting of sperm whales by other nations did occur relatively close to the Northwestern Hawaiian Islands. Between 1973 and 1976, over 1,900 sperm whales were killed by Soviet and Japanese whalers within 200 nautical miles of the Hawaiian Islands, almost all to the north of the Northwestern Hawaiian Islands. While sperm whales are no longer the subject of commercial whaling in the North Pacific, they face other threats. Japanese "scientific whaling" in the North Pacific does include small numbers of sperm whales; in recent years, a couple of animals have been killed each year. In high-latitude areas such as off Alaska, sperm whales do take fish off longlines and sometimes get hooked as a result. In recent years, only one sperm whale has been documented being hooked in the offshore longline fishery in Hawai'i; thus such interactions appear to be rare and likely do not jeopardize the population. Taking fish off lines is a learned behavior and can spread throughout populations, so it is possible that this type of behavior could become more common in the waters of the central Pacific. Because sperm whales range widely—especially adult males that spend a substantial proportion of their time in high-latitude areas—it is possible the population that uses Hawaiian waters could be impacted by interactions with fisheries elsewhere, including areas off Alaska. Given their behavior of logging at the surface, they are occasionally hit by vessels, and with a diet primarily of squid, they are one of the species that at least occasionally ends up mistakenly consuming plastics, which can lead to impaction and death. It seems that overall, however, none of these threats are likely to prevent sperm whale populations from increasing, although surprisingly there is very little evidence from anywhere that sperm whale populations have rebounded since the end of whaling.

THE BALEEN WHALES

About 25 to 35 million years ago, the whales split into two distinct groups: the toothed whales and the baleen whales. Baleen hangs in plates from the roof of the upper jaw and is used to filter prey out of the water. It is made up of keratin, the same material that makes up hair or fingernails. Without actually seeing a view of the baleen or teeth, the most obvious anatomic difference between these two groups are the blowholes. Baleen whales have a paired blowhole, while toothed whales have only a single blowhole. Most of the baleen whales are relatively large,

Most of the baleen whales are gulp feeders, with throat grooves or pleats that expand, allowing the whale to take in a large mouthful of water and prey. This is a Bryde's whale off Mexico, with a sea lion visible in the distance to the left. Photo © Doug Perrine.

but there is a lot of overlap in size between toothed and baleen whales, so size alone doesn't help much in identifying the type of whale.

When people think about whales in Hawai'i, they almost always think about humpback whales. Hundreds of thousands of tourists who come to Hawai'i each winter end up on boat tours to see humpbacks or see them from shore. The whales leave their summering grounds off Alaska and migrate to the nearshore waters around the Hawaiian Islands from late December through early April. Whale watching can be done either from shore or from tour boats, and there are commercial operators that take people out off Maui, O'ahu, Kaua'i, and Hawai'i Island. But humpback whales are not the only baleen whale to inhabit these waters. There are actually six other species of baleen whales that have been recorded in Hawaiian waters, including the largest of all whales, the blue whale, and one of the smallest of the baleen whales, the common minke whale.

Two different families of baleen whales have been documented in Hawaiian waters. There is a single member of the family Balaenidae, the North Pacific right whale, that has been seen only a handful of times over the last fifty years. Right whales have a highly arched jaw and lack the throat grooves found on many baleen whales. In Hawaiian waters there are six members of the family Balaenopteridae, commonly called rorquals, which include the familiar humpback whales, as well as the more streamlined blue whales, fin whales, sei whales, Bryde's whales, and common minke whales. The rorquals have a much straighter jawline and have varying numbers of throat grooves or ventral pleats that allow the throat to expand greatly when they are feeding, allowing them to engulf huge quantities of prey and water at once. All but one of the rorquals in the central North Pacific are widely distributed and seasonal migrants. In general, they spend their summers in high-latitude waters feeding and move to tropical areas, including Hawaiian waters, for part of the winter, primarily to find mates or give birth. But unlike humpback whales, which congregate in large numbers in shallow nearshore waters during the winter, these other species—blue, fin, sei, and minke whales—tend to spend their time in deep waters far from shore and are typically seen alone or in very small groups. Their tendency of spending most of their time far from shore has resulted in relatively few sightings and low population estimates of most species, particularly because the large-vessel offshore surveys undertaken by NMFS in 2002 and 2010 were in the summer and fall rather than

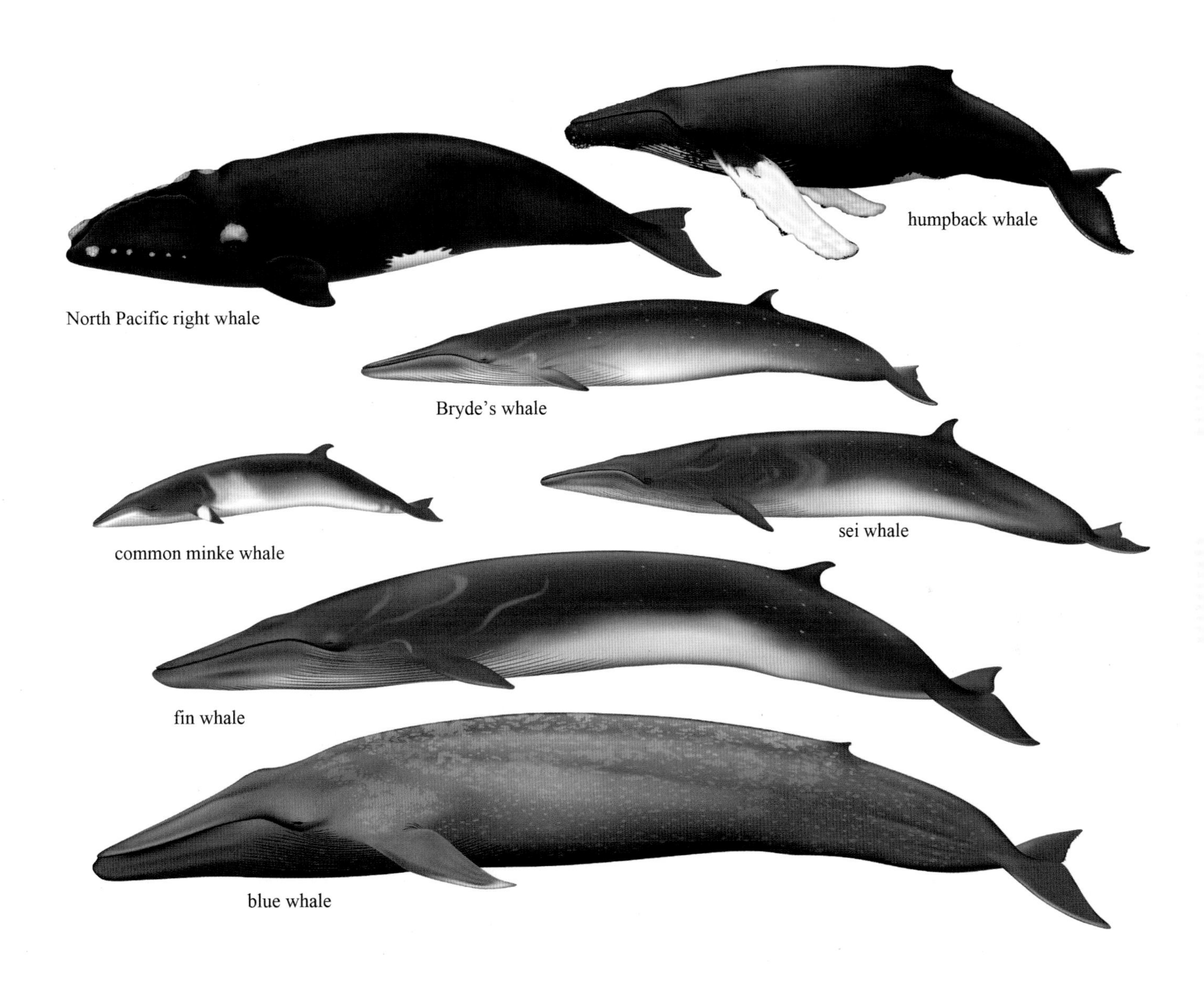

Seven species of baleen whales have been recorded in Hawaiian waters: North Pacific right whales, humpback whales, minke whales, Bryde's whales, sei whales, fin whales, and blue whales. Illustrations by Uko Gorter.

in the winter, when these species are most abundant in Hawaiian waters. Two of the smaller balaenopterids, the Bryde's and Omura's whales, stay in the tropics and subtropics year-round, but only the Bryde's whale has been recorded in Hawaiian waters.

All of these baleen whales have been hunted in the North Pacific. Hunting of the slower species, North Pacific right whales and humpback whales, began in the nineteenth century, while the faster species were not killed in earnest until the twentieth century, when whaling vessels were capable of greater speeds. The number of baleen whales killed in the North Pacific in the twentieth century was an astonishing 237,000 individuals, ranging from 967 right whales to 75,538 fin whales. Most of these were killed on their northern feeding grounds, where they could be predictably found in higher densities than in tropical areas. Five of the seven species of baleen whales that have been documented in Hawaiian waters—all but Bryde's and minke whales—are currently listed as endangered under the Endangered Species Act, all due to declines from large-scale commercial whaling as well as illegal Soviet whaling.

NORTH PACIFIC RIGHT WHALES *(Eubalaena japonica)*

Right whale sightings are so rare in Hawaiian waters that two peer-reviewed papers were published based on a single sighting off Maui on March 25, 1979. Right whales were given this name as they were the "right whale" to hunt: they are slow moving and float when they are killed, and their thick blubber layers provided huge quantities of oil that were sold for tremendous profits. Hunting of North Pacific right whales began in 1835, and only fourteen years later the population had been reduced to such an extent that whalers switched to other species. Both of the northern hemisphere species, North Pacific right whales and North Atlantic right whales, were hunted close to extinction. It is thought that there are two distinct populations of North Pacific right whales: a population of several hundred individuals in the western North Pacific centered around the Sea of Okhostk and a population of just tens of individuals in the eastern North Pacific found primarily in the Bering Sea and Gulf of Alaska. Unlike the North Atlantic right whales, which are known to breed in shallow nearshore waters along the southeastern U.S. coast, North Pacific right whales are thought to breed elsewhere, presumably in open-ocean waters, but this is still a mystery. This species has been recorded in Hawaiian waters only on a handful of occasions, and the last confirmed sighting appears to have been in 1996 off the west side of Maui. Right whales were probably never commonly found in Hawaiian waters. When nineteenth-century whalers first started using Lahaina and other ports in Hawai'i, there were no reports of right whales being hunted locally, despite their strong commercial importance at that time.

Right whales are one of the largest baleen whales, reaching a length of at least 17 m (55′9″) and weighing up to 90,000 kg. From a distance, the most likely way of recognizing a right whale is their distinct V-shaped blow, reaching up to 5 m in height. Their broad and relatively flat back, with no hint of a dorsal fin, is unlike any other species of whale in Hawai'i. Their head is massive, almost a third of their total length, and their jawline is strongly curved, making room for baleen plates that can be up to 3 m long. Right whales are skim feeders—they eat copepods and other zooplankton, swimming with the mouth open through patches of prey. They regularly lift their tail flukes out of the water when they dive, one of only three species of large whales in Hawai'i that regularly do so (the

A North Pacific right whale breaching in Alaskan waters. Photo by Brenda K. Rone/NMML.

[Top] The distinctive V-shaped blow of a North Pacific right whale in Alaskan waters. Photo by Amy R. Kennedy/NMML. [Bottom] The distinctive head of a North Pacific right whale, showing the callosities and the highly curved lower jaw. Photo by Brenda K. Rone/NMML.

other two being sperm whales and humpback whales; blue whales do so only occasionally). Combined with their all-black dorsal coloration and callosities (areas of roughened skin) on the head, they are likely to be easily recognized, except for one inconvenient fact: they are most likely to be seen in winter months, when large numbers of humpback whales are in the area, and they may be seen mixed in with humpbacks, making them easy to miss. The two individuals sighted in recent years, one in 1996 and one seen twice in 1979 (with the two sightings sixteen days apart), were both seen accompanying humpbacks.

Both of the right whales photographed in recent years were distinctive, and one of the two, the individual seen in 1996, was resighted and photographed 119 days later in the southeastern Bering Sea, suggesting that the right whales that visit Hawaiian waters are likely part of the eastern North Pacific population. The same individual was again photographed in the Bering Sea in 2008, 2009, and 2010.

Prospects for seeing a North Pacific right whale in Hawaiʻi are very low, and they are likely to continue to remain so. This species is one of the rarest whales in the world, with the eastern North Pacific population numbering just over thirty individuals, and the majority of those seem to be males. There is no sign that the population is recovering from their decimation by commercial whaling and illegal pelagic Soviet whaling in the 1960s. That said, of the baleen whales that use Hawaiian waters other than humpbacks, North Pacific right whales may be among the most likely to be in very shallow water, such as the Maui Nui Basin, so people should watch out for them.

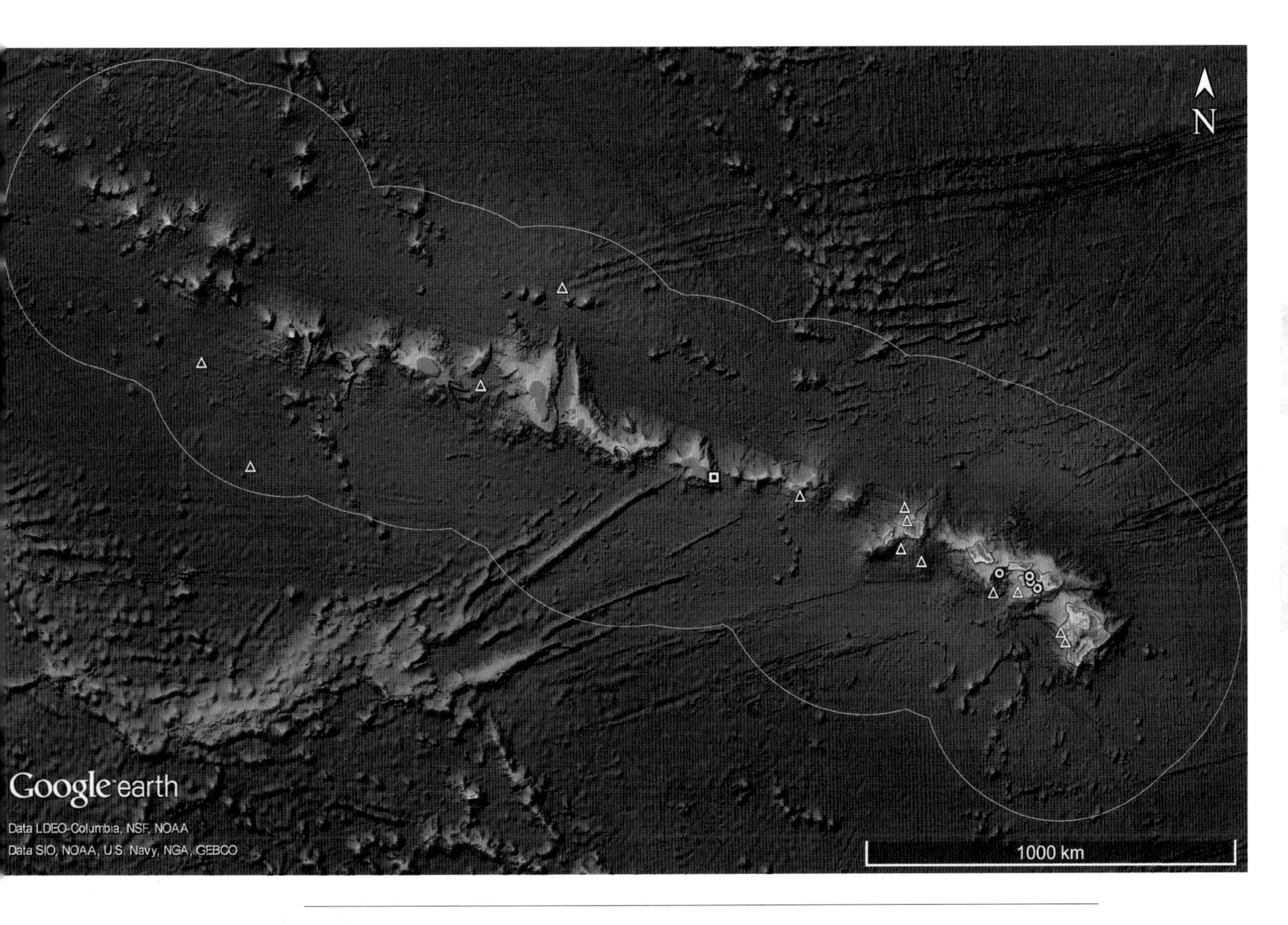

Sightings of North Pacific right whales (yellow circles), fin whales (white triangles), and a single sighting of two blue whales (white square). These include published records, sightings from NMFS surveys, and ongoing Cascadia research from 2000 through the end of 2015.

BLUE WHALES *(Balaenoptera musculus)*

The loud, low-frequency calls of blue whales, a moan that is below the hearing range of most humans, travel an incredible distance underwater, and some of the first evidence that blue whales used Hawaiian waters came from recordings from hydrophones offshore of Kāne'ohe, O'ahu, in the late 1970s. There were two different types of blue whale vocalizations recorded off Hawai'i: one type during the winter, when they would be expected to migrate toward the tropics from higher-latitude feeding areas, and another type recorded less often in the summer. More recent research has shown that these two blue whale calls represent two different populations of blue whales in the North Pacific—an eastern North Pacific population that overlaps with Hawaiian waters in the summer and a western and central North Pacific population that uses Hawaiian waters in the winter.

Blue whales are the largest of the whales. They are born at about 7 to 8 m long (~23′–26′3″) and in the North Pacific may reach lengths of over 24 m (~80′). Their blow is tall and straight, reaching 10 to 12 m high. Although one might think their large size and slate-gray or bluish coloration would make them easy to recognize, they overlap in size with fin whales and sei whales, so a large streamlined whale in the distance is not necessarily going to be a blue whale. Of the rorquals, blue whales have the smallest dorsal fin in proportion to their size, the most prominent "splash guard" in front of the blowholes, and a rounded head when viewed from above.

Blue whales feed primarily on krill, and while most of their feeding is thought to occur in colder waters, they will feed in productive areas anywhere. There is one feeding aggregation that occurs in the tropics, in an upwelling area off of Costa Rica, and another in the subtropics off Baja California, Mexico, where they sometimes feed on pelagic crabs.

Although they have been known to use Hawaiian waters for a long time, there are very few sightings of blue whales in Hawai'i, and all are to the north of the island chain. Observers on longline vessels have seen a few individuals north of the Northwestern Hawaiian Islands and well north of Maui, and during the 2010 NMFS survey there was a sighting of two blue whales associated with fin whales in the Northwestern Hawaiian Islands. From the 2010 NMFS survey, abundance was estimated at 133 individuals in Hawaiian waters during the summer and fall,

A mother and calf pair of blue whales off California, showing the small dorsal fin and large expanse of back. Photo by John Calambokidis.

One of the two blue whales documented during an NMFS research cruise in Hawai'i, off Necker Island on November 24, 2010. The mist from the blow of this individual can be seen more than 10 m high in the air. Photo by Jim Cotton/SWFSC.

although there was a lot of uncertainty associated with the estimate. This estimate was also from the time of the year when many of the blue whales that use Hawaiian waters were likely still feeding farther to the north, so numbers during the winter may be greater. We've never seen a blue whale in our work around the main Hawaiian Islands, and I'm not aware of any sightings near the main islands. However, there have been occasional acoustic detections of blue whales off the Kona Coast of Hawai'i Island from throughout the year, although the calls can be heard over a long distance, so the whale or whales may have been dozens of miles offshore. Blue whales are commonly seen in areas such as off southern California, and there is some evidence that the eastern North Pacific blue whale population has increased since the end of commercial whaling. Whether this will result in an increase in sightings in Hawaiian waters is unknown.

FIN WHALES *(Balaenoptera physalus)*

Although my work in Hawai'i started off the islands of Maui and Lāna'i, and we had projects there each year through 2003, it wasn't until December 2012 that we managed to get back to the area, with all our work between those dates off the other Hawaiian Islands. We based the 2012 project out of Mānele Bay on the south shore of Lāna'i, in order to be able to work in the deeper waters to the south and west of the island. After thirteen years of working in Hawai'i, we'd documented all eighteen species of odontocetes known to be here, but we had yet to see any species of baleen whale other than humpbacks. It was early December and humpbacks were in the area, and we were working with a dispersed group of pilot whales when I spotted a large blow about a mile away with pilot whales around it, so we headed over. Needless to say, we were all incredibly excited when a large baleen whale surfaced—and it wasn't a humpback! The pilot whales appeared to be harassing the whale, and it was moving quickly, but we were able to follow it.

Fin whales are the second longest of the whales, born at 6 to 6.5 m (19′8″ to 21′4″) and reaching lengths of close to 24 m (78′9″), and this animal was at least 18 m (almost 60′) long. One of the diagnostic features of a fin whale is the coloration of their lower jaw, dark on the left and white on the right, but it was difficult to see given the lighting conditions. The large dorsal fin rising off the back with a gentle slope made it look like a fin whale, but we collected a skin biopsy to confirm; genetic analyses confirmed this was our first sighting of a fin whale in Hawaiian waters. The distinctive scarring pattern from cookie-cutter shark bites and a hole through the fin, possibly also from a cookie-cutter shark, made the individual quite distinctive, and we will recognize it again if we get photos.

Whalers operating offshore of the Hawaiian Islands in the early 1800s regularly recorded "finback" whales, but they didn't discriminate between fin whales, sei whales, and Bryde's whales. Fin whales weren't documented in Hawai'i until the 1950s, based on a stranding on Maui. There are a lot more sightings of fin whales in Hawaiian waters than there are of blue whales. They've been seen in offshore waters during the 2002 and 2010 NMFS surveys, off the Northwestern Hawaiian Islands, and there are a handful of sightings around the main Hawaiian Islands. Our second sighting of fin whales was in February 2015—a

The head of a fin whale as it surfaces south of Kaua'i, February 12, 2015. This whale was one of a pair of adults. Photo by Brenda K. Rone.

A fin whale off of Lāna'i, December 7, 2012. This species can be distinguished from the other baleen whales by the gentle slope of the dorsal fin as it rises up from the back. Photo by author.

pair of individuals south of Kaua'i traveling north toward the Kaulakahi Channel between Kaua'i and Ni'ihau. They've been seen occasionally north of O'ahu and Kaua'i, and a pair of individuals was seen off Kona in January 2015. This sighting off Kona is among the southernmost sightings of fin whales in the central North Pacific—they appear to be largely absent in equatorial waters from about 20°N to 20°S in the Pacific. Fin whales do appear to primarily use Hawaiian waters in the winter, although there are acoustic detections from off O'ahu in ten months of the year, all except June and July. There are also occasional acoustic detections off Kona, although primarily in the fall and winter. The summer and fall estimate of abundance for all of Hawaiian waters from the 2010 NMFS survey was 154 individuals, but again there was a lot of uncertainty associated with that estimate, and numbers should be higher in the winter.

Fin whales have a slightly more diverse diet than blue whales, feeding on krill, small schooling fish such as herring and sand lance, and also occasionally on squid. More than seventy-five thousand fin whales were killed in commercial whaling operations in the North Pacific in the twentieth century, and they were protected in the North Pacific in 1976. Although nothing is known about population trends in Hawaiian waters, there is evidence for a fairly steady recovery of the fin whale population off the west coast of North America over the last twenty years, as well as an increase in Alaskan waters, suggesting that the population that uses Hawaiian waters is probably also increasing. If so, sightings of fin whales around the main Hawaiian Islands should continue to increase in frequency.

SEI WHALES *(Balaenoptera borealis)*

Sei whales (pronounced "say") were first documented in Hawaiian waters in March 1973 by Japanese whalers, with several animals marked with "Discovery" tags near French Frigate Shoals in the Northwestern Hawaiian Islands. Discovery tags were round metal cylinders with unique numbers, shot into the whales, so that if they were killed at a later date some information on their movements could be determined. Two of the individuals marked, both adult females, were later killed by Japanese whalers, one in August 1973 over 2,400 km (1,500 mi) to the north and the other in August 1974, over 2,000 km (1,300 mi) to the northeast. In June 1973, one sei whale was killed in far northern Hawaiian waters by Soviet whalers.

There have been only a dozen or so confirmed sightings of sei whales in Hawaiian waters since, including in offshore waters north and south of the island chain, in the Northwestern Hawaiian Islands, and off the main Hawaiian Islands. Sei whales are about as long as an adult humpback, but like the rest of the rorquals they are much more streamlined. They were heavily hunted in the North Pacific, with almost seventy-four thousand individuals killed in commercial whaling operations in the twentieth century, and the population was thought to have been reduced to about 20 percent of the pre-whaling numbers. Sei whale sightings are rare throughout their range in the North Pacific for a couple of different reasons. They tend to be found farther offshore than most of the other rorquals, and they are difficult to distinguish from other species.

At birth, sei whales are about 4.5 to 4.8 m (14′9″ to 15′9″) long, and adults are typically 12 to 17 m (39′4″ to 55′9″) long. Like fin and blue whales, they have just a single ridge running down the center of the head, while Bryde's whales (pronounced "brewdus") have three ridges, a central one and two smaller "accessory" ridges. There are a couple of different ways to discriminate fin and sei whales; the easiest is the angle that the dorsal fin rises up from the back. The dorsal fin of a sei whale rises steeply from the back and then curves backward, while in fin whales it rises from the back gradually. The head of a sei whale is more curved, and the tip is downturned, unlike either fin whales or Bryde's whales. The coloration of the lower jaw of a sei whale is also symmetrical, although it is a lot easier to see the angle of the dorsal fin than it is to get a good view of the coloration of

A sei whale underwater off Kona, Hawai'i, in 1995. The diagnostic downward curving tip to the rostrum is visible. Photo © Doug Perrine.

[Top] A sei whale seen northeast of O'ahu, November 16, 2007. Only a single head ridge and the slightly downcurved tip of the rostrum is visible. Photo by Tom Jefferson. [Bottom] A sei whale seen November 7, 2010, far offshore to the south of the Northwestern Hawaiian Islands. The relatively flat surfacing profile typical of sei whales can be seen in this image, as well as the dorsal fin rising steeply from the back, similar to Bryde's whales and minke whales. Photo by Jim Cotton/SWFSC.

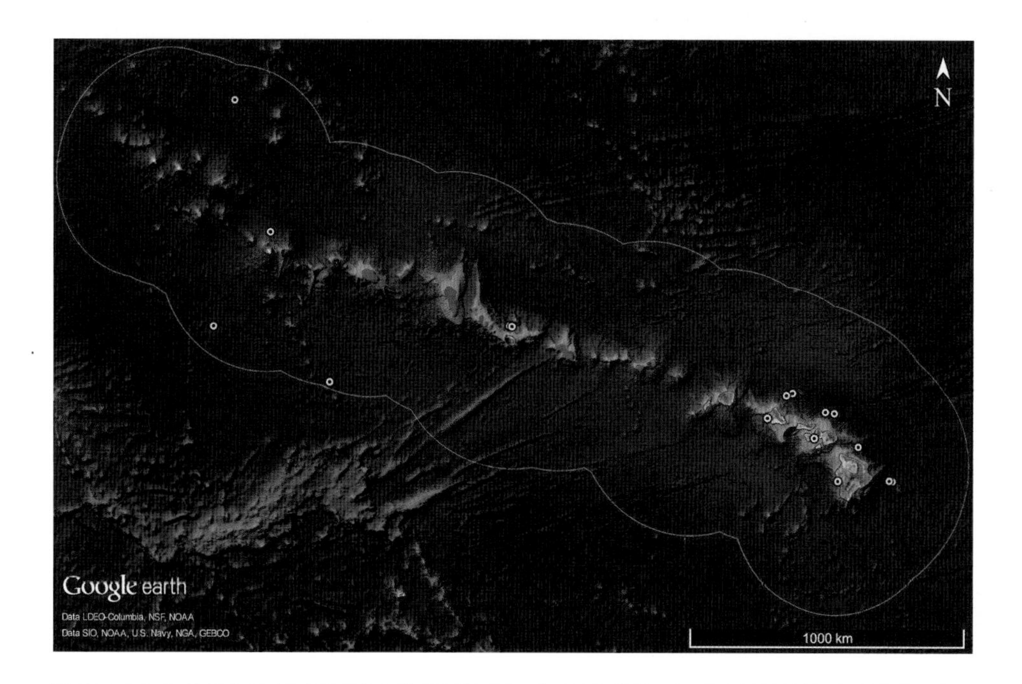

Records of sei whales in Hawaiian waters, from NMFS surveys, published records, individuals marked with "Discovery" tags by Japanese whalers, and recent unpublished records. The northernmost record shown is a single animal killed by Soviet whalers in June 1973.

the right-hand lower jaw. Sei whales also tend to surface differently, emerging at a much shallower angle than either fin or Bryde's whales, rarely arching their back. Telling sei and Bryde's whales apart can be challenging. A good view of the top of the head to see whether there is one (sei) or three (Bryde's) head ridges or the profile of the head from the side to see whether it is flat (Bryde's) or curved (sei) may be needed. The dorsal fins of both species rise steeply from the back, but sei whale fins tend to have a distinct backward bend about halfway up from the base. This difficulty in distinguishing them at sea is reflected in the abundance estimates from large-vessel surveys. Most surveys have three categories—sei whales, Bryde's whales, and sei or Bryde's whales—and the third estimate is often greater than either of the others.

Sei whales appear to be quite flexible in their diet, and they skim feed more than the other rorquals, feeding on copepods, krill, and small schooling fish such as sardines and anchovies. Sei whales are occasionally known to aggregate in groups of twenty to fifty individuals. Although this has not been confirmed in Hawaiian waters, there have been several large groups of baleen whales (that were not humpback whales) seen off Kona in recent years, and it is possible these may have been either sei or Bryde's whales. Abundance of sei whales in Hawaiian waters was estimated in the 2010 NMFS survey as about 390 individuals. As with the blue and fin whale estimates from this survey, in theory there may be more than that number during the middle of winter, when they are likely to be most common in Hawaiian waters. The most recent sighting I'm aware of was a pair of individuals seen off Wai'anae in January 2015. Unlike fin whales, there is no evidence that sei whale populations in the North Pacific are showing any signs of recovery, so sightings are likely to remain infrequent.

BRYDE'S WHALES *(Balaenoptera edeni)*

Bryde's whales were first documented in Hawaiian waters by Soviet whalers in May 1973. Between then and July 1976, Soviet and Japanese whalers killed a combined 307 individual Bryde's whales within 200 nautical miles of the Hawaiian Islands, almost all to the north of the Northwestern Hawaiian Islands. The first known sighting of a Bryde's whale in Hawaiian waters that wasn't killed was to the southeast of Nihoa in April 1977. Bryde's whales are the second most abundant of the balaenopterids in Hawai'i after humpback whales, and they are the only one found in Hawaiian waters year-round. They are restricted to the tropics and subtropics—the only species of rorqual in Hawaiian waters that does not have north-south seasonal migrations.

During the 2010 NMFS survey, there were thirty-two sightings of Bryde's whales, another twelve of sei or Bryde's whales, and eleven unidentified rorquals in comparison to just two sightings each of sei whales and fin whales and a single sighting each of blue whales and minke whales. The abundance of Bryde's whales in Hawaiian waters in the 2010 NMFS survey was estimated at 1,751 individuals, and there was another estimate of 766 sei or Bryde's whales, so Bryde's whales may be relatively common in Hawaiian waters. But what is surprising is that virtually all the sightings from the 2002 and 2010 NMFS surveys were either in offshore waters or around the Northwestern Hawaiian Islands. The only confirmed sightings near the main Hawaiian Islands

The head of a Bryde's whale photographed off Mexico, showing the central ridge and the pair of accessory ridges, one on either side. Photo © Doug Perrine.

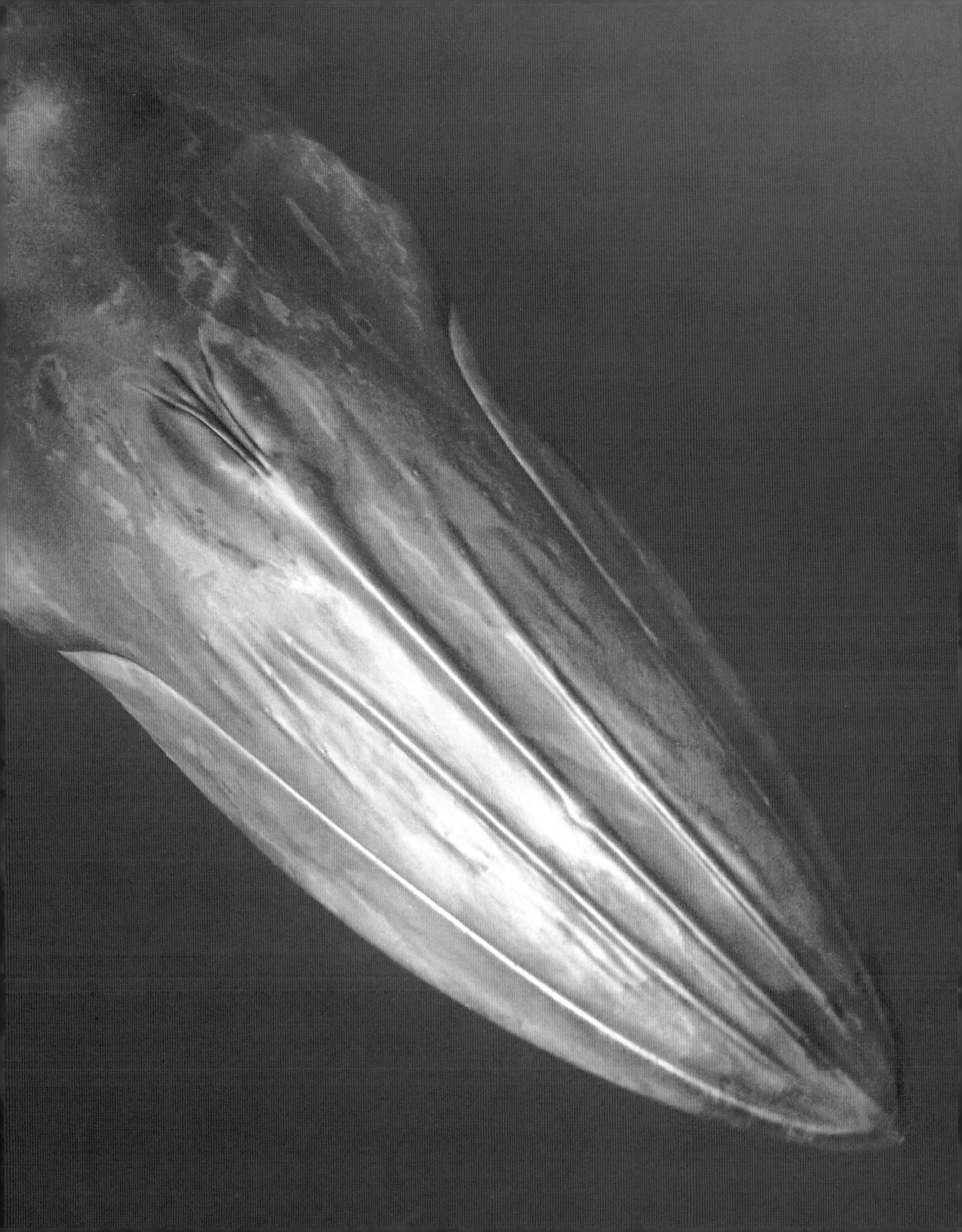

were one north of Niʻihau and another northwest of Molokaʻi. I'm not aware of any confirmed sightings from Hawaiian waters since the 2010 survey, although this mainly reflects the lack of offshore surveys since then. Bryde's whales are known to occasionally aggregate in groups of ten to twenty individuals.

Bryde's whales are about 4 m (13′1″) long at birth and may reach lengths of 15 to 16 m (49′3″ to 52′6″). In the past, a "pygmy Bryde's whale" was recognized, but this form was described as a separate relatively unrelated species in 2003—Omura's whale, found in the western Pacific and Indian Oceans.

Bryde's whales were hunted in the North Pacific, with almost fourteen thousand individuals killed in the twentieth century, but nothing is currently known about population trends. As they are nonmigratory, they likely feed in Hawaiian waters year-round. They tend to be gulp feeders, lunging through schools of fish and also eating squid, krill, or other invertebrates, although what they feed on in Hawaiian waters is not known.

A Bryde's whale seen northeast of O'ahu, November 13, 2007. The dorsal fin rises steeply from the back, and the back arches more than is typical for sei whales. Photo by Tom Jefferson.

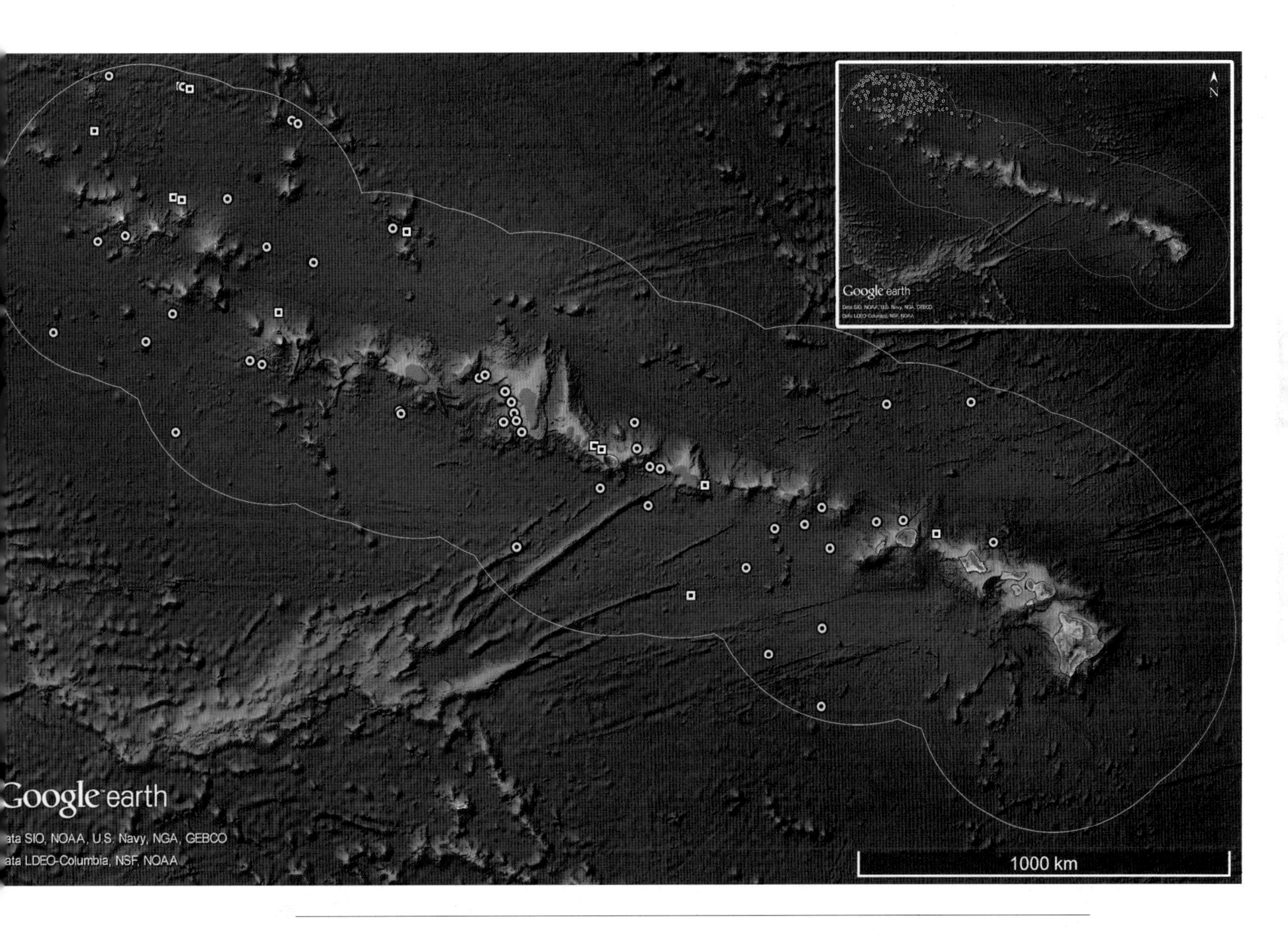

Sighting records of Bryde's whales (yellow circles) and whales that were either Bryde's or sei whales (white squares) within Hawaiian waters, from NMFS surveys and published records. The inset map shows locations of 307 Bryde's whales killed by Soviet and Japanese whalers betweeen 1973 and 1976.

COMMON MINKE WHALES *(Balaenoptera acutorostrata)*

In the 1950s, U.S. Navy submarines monitoring noise in the oceans off Hawai'i and southern California heard sounds that were later described as "boings." A paper describing whale sounds recorded off Kāne'ohe, O'ahu, in the late 1970s and early 1980s included boing sounds, even though it wasn't known what caused them. They showed up off O'ahu in December each year and were typically gone by May, a seasonal occurrence that was similar to what was recorded for humpback whale songs. They also found that when there was one sound source, there were long pauses between boings, whereas when there were many sources of the boings, they would chorus, like chorus frogs, suggesting that an animal was responsible for the boings and they were responding to others. Another study

The relatively pointed, small head of a minke whale, off Washington state. Photo by author.

looked at the distribution of boings around the Hawaiian Islands recorded from submarines and from hydrophones dropped off navy vessels or from airplanes. This study noted that boings showed up north of the islands in October through December, both north and south of the islands in January through March, and then moved back north again in April through June. The source of the boings remained a mystery until November 7, 2002, when the NMFS survey in Hawaiian waters that was combining acoustics and visual observations began tracking an animal that was making the boing sounds. Triangulating the sounds, they knew the source was close off the bow of the research vessel, and a minute after the sounds stopped a minke whale breached off the bow! After the whale dove, the boings started again.

Minkes are the smallest of the baleen whales in Hawai'i. Their blowholes are usually still visible when the dorsal fin breaks the surface. This photo was taken off Washington state. Photo by author.

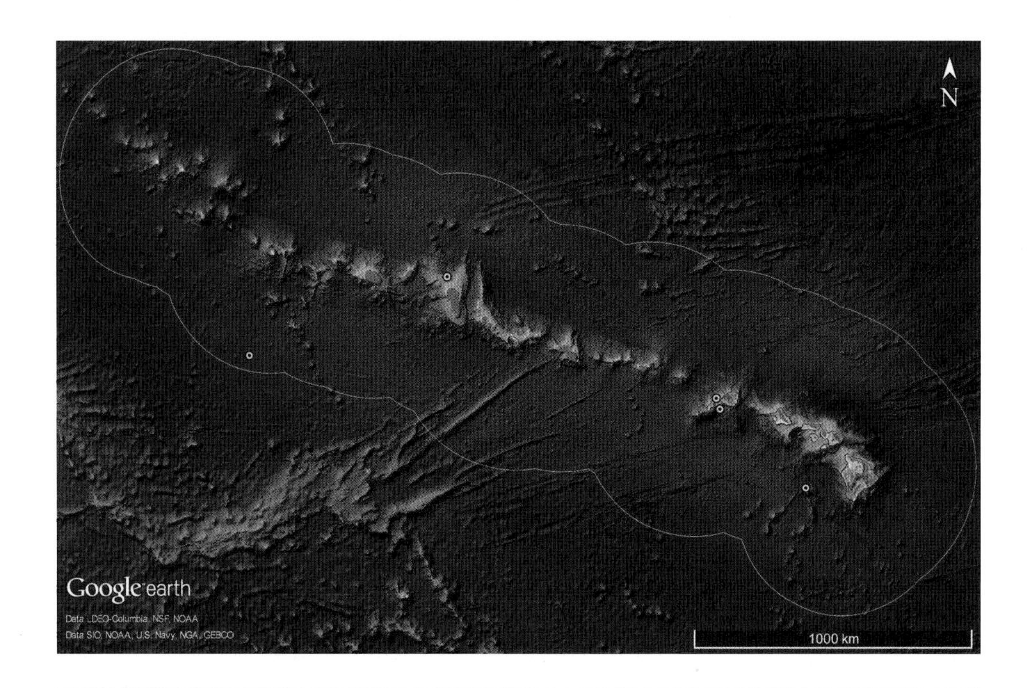

Although there are few sightings of minke whales in Hawai'i, their "boings" are regularly heard over hydrophones in winter months off northwest Kaua'i and north of O'ahu.

There are two species of minke whales recognized: the common minke whale, found throughout most of the world's oceans, and the Antarctic minke whale. Common minke whales are the smallest of the baleen whales in the North Pacific and North Atlantic. They are about 2 to 2.8 m (6′7″ to 9′2″) long at birth and may reach maximum lengths of about 8.8 m (28′10″). Their blow is usually not visible. They have a pointed head and a tall dorsal fin that rises steeply from the back. Although hard to see, the most distinctive feature is a white band on the flippers. At a distance, if just the back and dorsal fin are seen, a minke whale is likely to be confused with a Cuvier's beaked whale or a Bryde's whale, all having a similar arch to the back and a dorsal fin that rises steeply from the back. When minke whales surface, the dorsal fin is usually visible before the blowhole is submerged, unlike Bryde's whales.

Given how many minke whale boings have been detected in Hawaiian waters, there are surprisingly few sightings of them. In the 2002 and 2010 NMFS surveys, they were seen only twice, once off the Northwestern Hawaiian Islands and once far offshore to the west. There have been two confirmed sightings off Kaua'i, one in February 2005 and one in April 2009, and another sighting in January 2012 was likely a minke whale. Most of what is known about their presence in Hawaiian waters comes from acoustic studies. We detected boings offshore of Kona in April 2008, they've been detected northwest of O'ahu, and during the winter they are regularly detected northwest and north of Kaua'i. A year of recordings made in 2007–2008 at Station ALOHA, 100 km north of O'ahu, found that boings occurred from October until May, with a peak in March. Like most of the other baleen whales, minke whales migrate from high-latitude feeding areas to tropical areas to breed. Boings recorded in Hawai'i are a bit different from those recorded in California, suggesting there might be multiple populations of minke whales in the central and eastern Pacific. When in Hawaiian waters, minke whales are likely to be found alone or in pairs. Why there are so few sightings of minke whales, given the ubiquity of boings in the winter, probably reflects their use of waters off the northern sides of the islands, combined with their relatively small size and somewhat cryptic surfacing behavior.

HUMPBACK WHALES *(Megaptera novaeangliae)*

In the central North Pacific, humpback whales, *koholā* in Hawaiian, are a conservation success story: stop killing them, and the population rebounds. Over twenty-nine thousand humpback whales were killed in the North Pacific in the twentieth century. Humpback whales were hunted by all the Pacific Rim nations, including Taiwan, China, Japan, Korea, the Soviet Union, the United States, Canada, and Mexico. They were protected from commercial whaling in 1966 and listed as endangered in the United States in 1970. When research first started on humpback whales in Hawai'i in the mid-1970s, it was thought that there were perhaps a few hundred whales spending the winter here. From 1980 though 1983, the abundance of humpback whales using Hawaiian waters was estimated at about 1,400 individuals. Just over twenty years later, by the winter of 2005/2006, the estimate increased to about 10,100 individuals. In theory, in the eleven years since then the population may have continued to increase at about 5 to 6 percent per year, and thus by the winter of 2016/2017 the abundance of humpback whales using Hawaiian waters should be somewhere between 17,000 and 19,000 individuals. Although still listed as "endangered" in early 2016, they are no longer truly endangered, as the population continues to rebound, and NMFS proposed removing them from the endangered species list in early 2015. Why have they done so well compared to the other species of whales that were heavily hunted? They were one of the first species protected by the International Whaling Commission in the mid-1960s, and thus they have had more time to recover than many species. The resilience of humpback whales also likely reflects their adaptability to feed on many different types of prey. They may have the most diverse diet of any baleen whale, and they adopt a variety of different tactics to capture prey, allowing them to take advantage of opportunities as they arise.

A humpback whale breaching off Kona, Hawai'i, January 13, 2015, showing the long pectoral flippers that inspired the name of the genus *Megaptera*, for "large winged." Photo by Julie E. Steelman.

Identifying Features and Similar Species

Humpback whales are in a genus all to themselves, *Megaptera,* roughly translated as "large winged"—they have the longest flippers of any species of whale or dolphin. They are probably the most recognizable species of baleen whale, with a low, humped dorsal fin and flippers reaching about a third of their body length. The flippers range from all black to almost all white and are often raised out of the water while on their breeding grounds. There is a ridge that runs down the middle of the head with a series of tubercules on the ridge; they also have tubercles around the front and sides of the head. They raise their flukes out of the

The distinctive tail fluke of a humpback whale off Kona, Hawai'i. The markings on the underside of the flukes are used to identify individuals and thereby track their movements. Photo © Dan J. McSweeney/Wild Whale Research Foundation.

water when they dive more so than any other species of baleen whale in Hawai'i. The flukes have a serrated edge, unlike the generally smooth trailing edge of the flukes of other rorquals, with varying amounts of white on the underside of the flukes. Otherwise their body is largely dark gray or black. They are about 4.3 m (13′1″) long at birth. Adult females are typically 1 to 1.5 m longer than males, and females in the North Pacific may reach lengths of about 15 to 16 m (49′3″ to 52′6″). The blow is typically low and bushy, reaching about 3 m high. They are also more acrobatic than any of the other large whales in Hawai'i, regularly breaching and slapping their flippers or flukes on the surface.

Other species of large whales so rarely come into shallow waters in Hawai'i that almost any large whale near shore is likely to be a humpback, but offshore the species most likely to be confused with humpback whales in Hawai'i are sperm whales and fin whales. The bushy blow of a humpback whale rises vertically from the center of the head, unlike the oblique angle of a sperm whale blow, coming from the front left corner of the head. From a distance, humpbacks and fin whales might be confused, but the difference in dorsal fin shape should be obvious with a good view.

Habitat Use and Movements

Humpback whales spend much of the year in high-latitude feeding areas, and most individuals migrate to tropical areas for breeding. In the North Pacific feeding areas, they tend to hug the coast from California north to Alaska and across the western Pacific to Hokkaido, Japan. It is thought that about half the humpbacks in the North Pacific migrate to Hawaiian waters to breed and give birth. Other populations give birth in tropical waters in the western and eastern North Pacific.

Humpbacks were one of the first species of whales recognized to have individually distinctive markings. Individuals have been tracked over more than thirty years using the pigmentation patterns on the underside of the flukes, as well as the serrated trailing edge of the flukes, with identification photos obtained from behind the whales as they raise their flukes on a dive. One of the largest collaborative studies of whale movements and population structure in the world was the "SPLASH" study (Structure of Populations, Levels of Abundance, and Status of Humpbacks) in the North Pacific, with research undertaken in all the known breeding areas in the North Pacific over three winters (2004 to 2006) and in all the known feeding areas over two summers (2004 and 2005). From this work, it is clear that almost all of the humpbacks in Hawai'i come from a broad swath of the west coast of North America, ranging from northern Washington State to the Aleutian Islands and the Bering Sea. About 75 percent of the whales from southeastern Alaska and northern British Columbia head to Hawai'i for the winter, while about 50 percent of those in the Gulf of Alaska and the Aleutians do so as well. Movements from other areas in the North Pacific are rare, but individuals have been documented moving between Hawai'i and Mexico and between Hawai'i and Japan. Despite such movements, a comparison of the genetics of humpback whales in Hawai'i showed that they are distinct from other breeding populations to the west or the east.

How do they find the islands when crossing vast expanses of ocean? From examining the tracks taken by tagged humpbacks migrating north, it seems as if they use some geomagnetic cues to help in migrating, and it is also likely that they use low-frequency noises generated by the huge ocean swells as they break against Hawaiian shores. But why do they migrate to Hawai'i? It isn't for feeding—there is little evidence of feeding by humpback whales in Hawaiian waters, other than occasional snacking when a whale comes across a school of baitfish. For male humpback whales, it is obvious why they migrate: they are following the females. But the reason or reasons why female humpback whales migrate to tropical areas to breed are less clear. The main reason proposed is for females to give birth in an area with few predators that might harm the calves—in particular, killer whales, which are much more abundant in high-latitude areas. Certainly in Hawai'i, killer whales are rare. Humpback whales in Hawai'i show a strong preference for shallow waters, with most individuals in water less than a couple hundred meters deep, while killer whales in Hawai'i tend to be in much deeper waters. So nearshore Hawaiian waters are likely a good refuge from killer whales. Humpback whale flukes are often scarred by attacks from killer whales, but individuals that breed in Hawaiian waters have a much lower incidence of killer whale scarring than those that breed off Mexico. Given the infrequency of sightings of killer whales immediately around the main Hawaiian Islands, it is likely that most of the attacks on calves occur as the females and calves migrate north to the

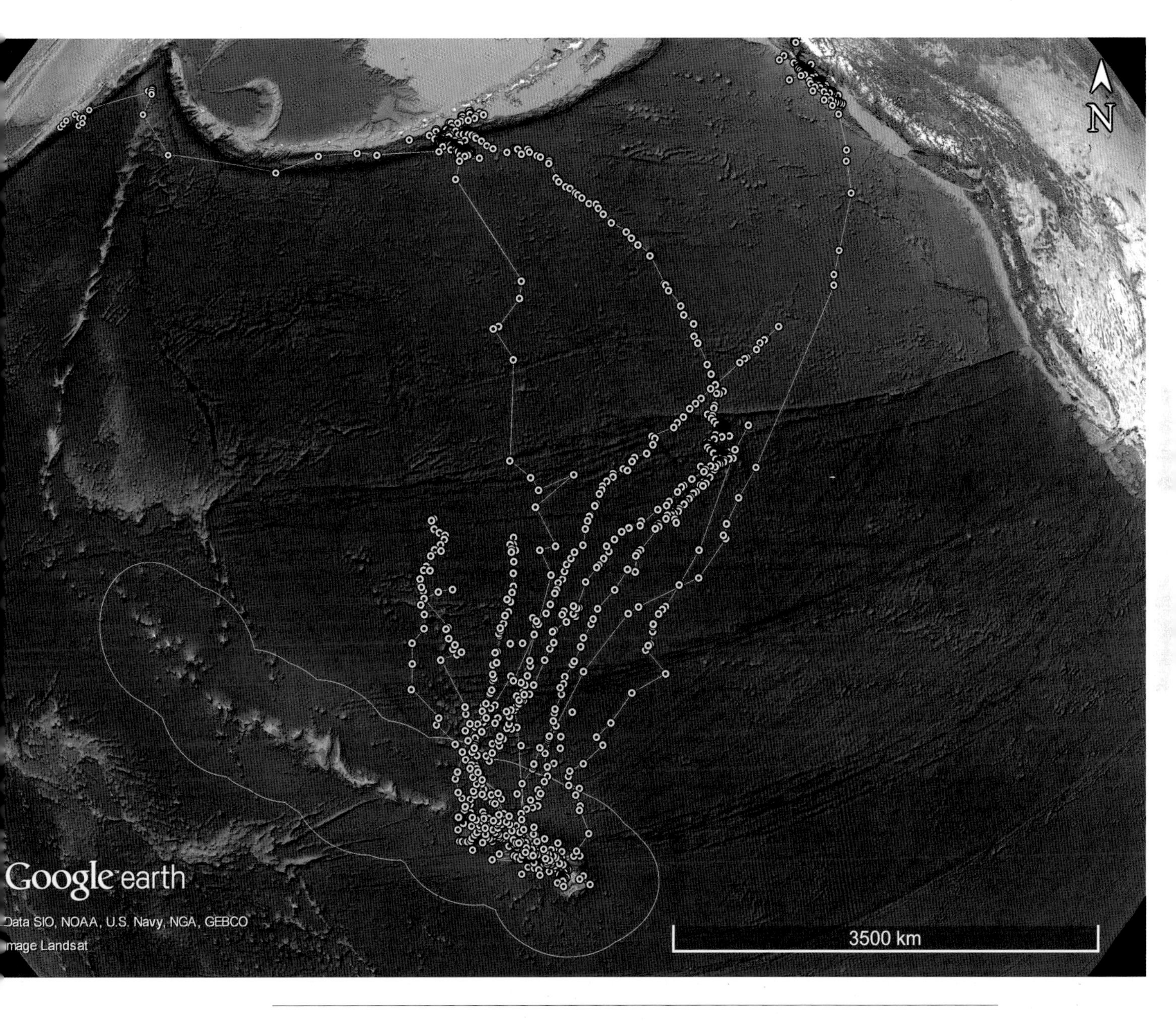

Locations from humpback whales satellite tagged in Hawai'i from 1995 through 2000, showing movements both among islands and migratory movements to southeastern Alaska, the Aleutians, and eastern Russia. Data from Bruce Mate, Oregon State University Marine Mammal Institute.

feeding grounds. A second hypothesis is based on strategic energy use; the warm waters allow for the offspring to conserve energy that can be devoted to growth.

During the SPLASH study, movements of individuals were regularly documented among the islands. Although most of the whales were identified off Maui, some of those were later documented off all the other study areas: Kaua'i, O'ahu, Penguin Bank, Moloka'i, and Hawai'i Island. A great deal is known about humpback whales in the main Hawaiian Islands. There is also evidence, however, that the Northwestern Hawaiian Islands are an important wintering habitat for humpbacks. The amount of shallow-water habitat available for humpbacks in the Northwestern Hawaiian Islands is almost twice that found in the main Hawaiian Islands. Acoustic recorders at a number of locations throughout the Northwestern Hawaiian Islands have found that humpback songs are very prevalent at Maro Reef, about halfway up the chain, and also at French Frigate Shoals and Lisianski Islands, similar to the levels recorded off O'ahu. Whether these are the same whales that use both areas is not known, but it is possible that the humpbacks using the Northwestern Hawaiian Islands represent a different breeding population entirely.

Life History and Behavior

Female humpback whales have their first calf at between five and ten years of age, and they come to Hawaiian waters both to give birth and to mate. For the other rorquals, courtship seems to be a relatively tame affair, with one male following a female around or perhaps calling to attract her attention, but with humpbacks, courtship seems to take a variety of roles, and things can get a bit more heated. Humpback whales are perhaps best known for their long and complex songs. Males will hang motionless in the water about 20 m down, head down and tail toward the surface, moaning and trilling for up to thirty minutes, repeating a song with themes and phrases, all sung in a particular order. All the whales on a breeding ground tend to sing the same song, and songs change over the breeding season. The next breeding season, the whales tend to pick up where they left off, and the song again morphs throughout the season. Male humpbacks are thought to sing for a variety of reasons: primarily to attract females, but songs are loud and carry a long distance, and other males obviously pay attention to their neighbors singing as well. On the Hawaiian breeding grounds, the density of male humpbacks is high, and females have a lot of choices, and whether a male sings or tries another strategy to get the attentions of a female may depend on his size or age.

In Hawaiian waters, many lone humpback whales may be "singers," or males otherwise looking for a female. Often pairs of adults are seen, and these are probably male-female pairs, with the male escorting the female, hoping for an opportunity to mate. Although females typically have one calf every two years, at times they will give birth two years in a row, so females with small calves may become pregnant. This is obvious to the males, as females with small calves often

A humpback whale lunging at the surface off Kona, February 21, 2015. Photo by Julie E. Steelman.

are accompanied by an escort. The most heated interactions are when a female is actively in estrous, and many males will compete with each other over access to the female. These are called competitive groups, typically with the female in the lead, a "primary escort"—often the largest male—close behind and a bit to the side, and up to dozen or more other males chasing behind, often challenging the primary escort. The number of escorts depends in part on the size of the female: larger females have more escorts, and larger females have larger calves, so they may be more desirable to males, attracting more competition. Whales lunge through the surface and ram each other, and both the primary escort and challengers will use their large and powerful tails to try to displace the other. Tubercles on the head are often left bloody after such interactions, dorsal fins may be injured, and individuals may even be killed on occasion or be left bloody and badly injured to the point where large sharks are attracted to the injured male and they may be unable to fend them off.

Humpbacks breach and tail lob (slap their tail flukes) more often than any of the other large whales in Hawai'i, and a lot of this behavior may be related to mating. Some of the percussive behaviors may be females trying to attract males to join in a competitive group to find out who the largest or most dominant one is. Other times such behaviors are by males, directed at other males or perhaps in frustration when they aren't the ones chosen. And finally, a lot of these types of behaviors are by juveniles, a form of exercise or building stamina for the long migration ahead.

The first independent funding I received in Hawai'i was a small grant from the Hawaiian Islands Humpback Whale National Marine Sanctuary to study humpback whale diving and nighttime behavior. In February, March, and April 2000, we deployed fifteen suction-cup-attached time-depth recorders on humpbacks off Maui, recovered fourteen of the tags, and got data back from thirteen of them—sixty-two hours of dive data, the first view into what humpbacks in Hawai'i were doing at night and below the depths that they could easily be followed by snorkelers. The whales we tagged that year were all thought to be males, most in competitive groups, but we also tagged one individual in a pair and the escort with a mother and calf. How deep they dove and their patterns of diving varied among the different types of groups. Some of the dives were to the bottom or close to it, in depths from about 100 to 170 m. There were no obvious

A humpback whale and a short-finned pilot whale off Kaua'i, February 12, 2015. In this encounter, several humpback whales were following and interacting with a group of pilot whales. Photo by Brenda K. Rone.

differences in diving behavior between day and night, which makes sense given that the whales are not thought to be feeding in Hawaiian waters and most of their activities are driven by sound rather than sight. Some of the tags included a paddle-wheel swim speed sensor, so we could tell when whales went motionless, possibly singing, for fifteen to seventeen minutes at a time. The whales were motionless at about 20 m depth; at that depth they were neutrally buoyant, neither sinking nor floating up to the surface. Diving patterns of whales in competitive groups were quite variable, but many of the tagged whales dove below 100 m depth, presumably because they were following the female. Why she was diving deep is unknown, but it could be to test the males—it is always tough to figure out how to choose the best one.

A humpback whale calf spyhopping off Maui. Photo by Andy Day/Cetos Research.

Humpback whales in Hawai'i often associate with other species, including other baleen whales. North Pacific right whales and fin whales have both been seen with humpbacks in Hawai'i. Interactions with various species of toothed whales and dolphins occur quite frequently. In shallow water, humpbacks overlap with spinner, spotted, and bottlenose dolphins, as well as false killer whales, but it is primarily bottlenose dolphins that interact the most with humpbacks. Bottlenose will often harass humpback whales, with the frustration of the humpbacks evident by the "trumpeting" or forced blows that they normally make in aggressive interactions. Bottlenose dolphins will also ride the bow wave created when humpbacks surface quickly. There are other times when the two species will play together, with humpbacks even lifting dolphins out of the water on the top of their heads. We regularly see humpbacks in the deeper Kaulakahi Channel between Kaua'i and Ni'ihau associating with short-finned pilot whales or rough-toothed dolphins. Typically these appear to be smaller humpbacks—juveniles or subadults—and they often seem to be the ones instigating the interactions, following pilot whales around rather than vice versa.

Conservation

Every year in Hawai'i there are individual humpback whales injured by vessel strikes, often by fast-moving whale-watching boats. Individuals are also at least occasionally entangled in fishing gear, including trap gear set in Hawai'i, although most of the entangled humpback whales in Hawai'i seem to be carrying gear brought down from feeding grounds in Alaska. An active humpback whale disentanglement network is run by the Sanctuary, and they are able to respond to and free many of the individuals reported entangled in Hawaiian waters. But despite these issues for individual whales, the population has shown a steady increase over the last thirty years, suggesting that the problems are not affecting the population. It seems likely that humpback whales will continue to increase in Hawaiian waters, at least as long as their food supplies to the north continue to be available. Humpbacks, along with fin whales, may be one of those species that benefits from climate change, as decreasing sea ice in the Arctic may actually increase foraging opportunities for humpback whales there.

The most endangered species of marine mammal in Hawai'i is actually the Hawaiian monk seal, with only about 1,200 individuals remaining for the entire species, while for whales and dolphins, it is typically one or more populations of a species that are in jeopardy from human activities in Hawai'i. Photo by Julie E. Steelman.

CONSERVATION OF HAWAI'I'S WHALES AND DOLPHINS

Conservation requires identifying which species are most vulnerable, what the main threats are, and then taking action to reduce or mitigate those threats. For Hawaiian whales and dolphins, there has been much progress toward the first two of these objectives but relatively little toward the third. For those species whose populations were greatly reduced due to commercial whaling, as well as illegal Soviet whaling—the larger baleen whales, as well as sperm whales—one would think that all we had to do was stop killing them and these species would have recovered. In the case of humpback whales, that has been true: since the moratorium on whaling, their population in the central North Pacific has rebounded. In 2015, the National Marine Fisheries Service proposed removing this population, along with many other humpback whale populations, from the list of endangered species. Fin whales in the North Pacific also appear to be recovering. However, for other large whales that were also heavily hunted, in particular North Pacific right whales and sperm whales, there is little to no evidence of recovery. In the case of right whales, it is probably because the population was reduced below some critical level, and recovery may either never happen or may take much, much longer than expected. For right whales, other types of human impacts such as ship strikes from large vessels or entanglement in fishing gear may hinder or prevent recovery. In the case of sperm whales, the lack of recovery may be due to whalers selectively removing the largest males from populations, to the point where pregnancy rates dropped—as in the eastern tropical Pacific, where there is evidence of pregnancy rates dropping after heavy whaling. Within groups of female and subadult sperm whales, the largest, and thus oldest, females were hunted first. The removal of the cultural knowledge of those matriarchs, as documented in elephants, may also be playing a role in the apparent lack of recovery. Both of these types of effects on sperm whales reflect the fact that social organization and cultural knowledge are important factors that influence the recovery of some species of long-lived social odontocetes. However, these species of large whales are wide ranging, and as far as we know they have no true Hawaiian populations. In this chapter I focus more on Hawaiian populations and how research results have played—or *should* play—a role in the management and conservation of these species.

According to recently revised estimates, the Hawaiian Islands were first colonized by Polynesians between about AD 1000 and 1200, and the human population probably peaked at a few hundred thousand people prior to the first documented visit by Europeans in 1778. There is little evidence, one way or the other, whether Native Hawaiians hunted nearshore whales or dolphins prior to European contact, although in the early 1800s there are reports describing hunting of dolphins off Maui and Oʻahu, and there is the earlier-mentioned account of a group of melon-headed whales driven into Hilo Bay and eaten. Certainly, their habits of resting near shore would have made spinner dolphins readily accessible to drive fisheries, although whether such hunting by Native Hawaiians occurred prior to the 1800s is unknown.

Spread of diseases brought by Europeans caused the Native Hawaiian population to plummet, and throughout the second half of the nineteenth century there were fewer than forty thousand natives in a total human population of less than one hundred thousand. Today the population of Hawai'i is close to 1.5 million people, a tenfold increase from the population in 1900. The population of Hawai'i since 1900 has increased faster than the global average and will probably continue to do so. With this many people, it is not surprising that we find an increase in human effects on whale and dolphin populations in Hawaiian waters. These include the effects of persistent organic pollutants (POPs) on the immune system, harassment of groups or individuals by well-meaning dolphin or whale watchers, and bycatch in fisheries, among others.

PERSISTENT ORGANIC POLLUTANTS

One of the conservation issues facing Hawaiian whales and dolphins arises not only from human activities in Hawai'i but as a by-product of global industrialization and the large and expanding global human population—the issue of POPs. Whales and dolphins are at the top of the food web. As they are long lived and have a layer of blubber where they store energy, they accumulate high levels of some lipophilic (fat-loving) pollutants that can compromise their immune system and/or affect reproductive rates. This is a difficult problem to recognize and to solve, as both the pollutants and their effects are insidious. POPs are not like an oil spill, where thousands of gallons of oil cover the water's surface. The pollutants that may cause immunosuppression—industrial chemicals such as PCBs, as well as pesticides and flame retardants used in fabrics and plastics—are dispersed throughout the environment, transported in ocean currents and even on dust particles in the atmosphere.

While PCBs were banned in the United States and much of the rest of the world in 1979, they resist environmental breakdown, some taking as long as a century to degrade, and like pesticides and flame retardants they accumulate in the blubber of whales and dolphins. These contaminants bioaccumulate, meaning that those animals at the top of the food web accrue much higher levels of contaminants, as their prey—predatory fish or mammals—have accumulated high levels before they became food for a higher-level predator. So, among the whales and dolphins in

Hawaiian waters, species least at risk are those that feed on small squid and deep-water fish, such as spinner dolphins and melon-headed whales, while those that feed on large game fish or other long-lived predators, such as false killer whales and killer whales, have the highest levels. Although individuals from the resident Hawaiian population of false killer whales do have high levels of mirex, a pesticide historically used in pineapple cultivation, they also have high levels of PCBs, most of which probably came from sources in Asia or elsewhere in the world.

There are no local solutions to solving the problem of POPs, as many of these pollutants travel throughout the world's oceans and atmosphere. The only real long-term solution involves international agreements and international action. The United States has been actively involved in these issues on the international stage, but there is still a need to ensure that global pollution concerns are part of new international agreements such as trade deals, for example. But local actions and choices by individuals can also play a role in minimizing the amount of such pollutants being added to the environment. We can encourage organic agriculture by buying organic products when possible, we can minimize the use of herbicides and pesticides in our yards and neighborhoods, and we can choose furniture or building materials (where there is a choice) that have received little or no treatment with chemicals such as flame retardants. Recognizing that these pollutants all originate on land and most enter the ocean through runoff and storm drains, actions to minimize their use and contain fallout when used are critical.

EFFECTS OF SOUND ON MARINE MAMMALS

Sound is another by-product form of pollution in the marine environment. Whales and dolphins use sound to find mates, avoid predators, maintain their social groups, and, in the case of odontocetes, to find food. If they are loud enough, sounds that humans introduce into the water have the potential to mask the use of sounds by whales and dolphins, reducing their ability to communicate, making it harder to find prey, or both. Noise levels in the oceans have increased dramatically over the last few decades, with a huge increase in shipping traffic as one of the main culprits. As an island state with almost 1.5 million people, not counting tourists, Hawai'i depends on large vessels to supply many of the goods and products needed to sustain the population. The routes these vessels take to

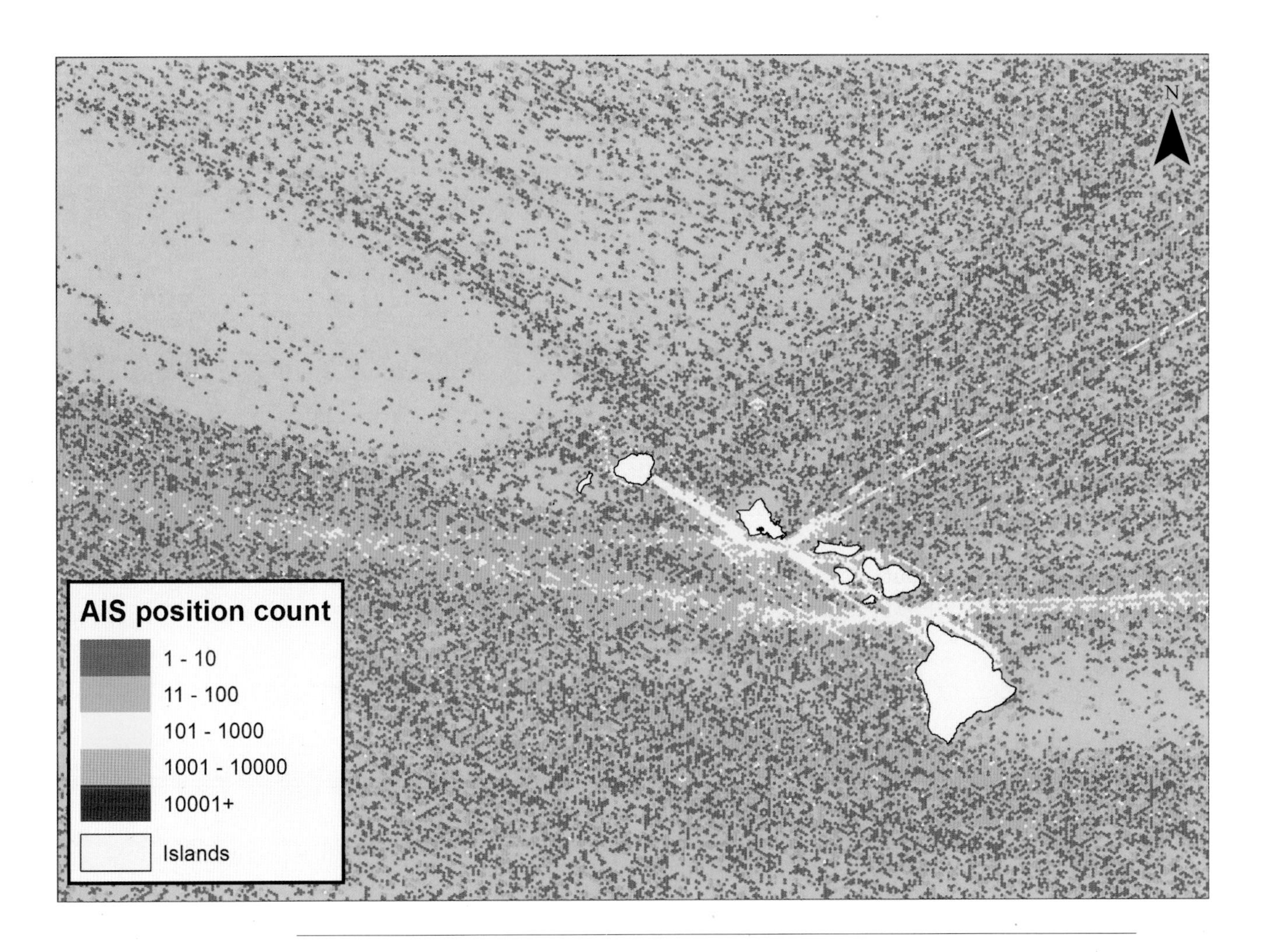

Vessel traffic is monitored remotely using an Automatic Identification System (AIS) on board all large and many small vessels. This map shows the major shipping routes to and from Hawai'i, with a concentration of vessel traffic off southern O'ahu, as well as the routes used among the islands. The large relatively sparse area to the northwest is within the Papahānaumokuākea Marine National Monument.

and from the islands are somewhat predictable and concentrated in certain areas, so any noise impacts, in terms of masking animals' hearing, are fairly predictable as well. Maps of shipping traffic in Hawaiian waters show that, not surprisingly, the south shore of O'ahu has the highest levels of large vessel traffic, with other high use areas in a band between the north end of Hawai'i Island and Nāwiliwili Harbor on Kaua'i, reflecting the main ports used for shipping among the islands. While the higher-intensity low-frequency sounds of large vessels may travel a great distance underwater, to some degree these sounds are transitory, and they overlap primarily with the hearing of baleen whales. The greatest acoustic footprint around resident whales and dolphins in Hawai'i is probably small vessel traffic, primarily tour vessels near shore and fishing vessels farther from shore, particularly when the vessels seek out different species of whales and dolphins.

The U.S. Navy produces the loudest human-created underwater sounds in Hawai'i, either using underwater explosives or high-intensity mid-frequency sonar. Naval sonar is one of many factors that may affect whale and dolphin populations in Hawai'i, although it is difficult to state with certainty what effects have occurred. A little bit of background on the types of sonar in use is important to an understanding of potential impacts.

There are many different types of sonar, varying both in frequency (from low to very high pitched sounds) and in intensity (how loud they are). The potential for an effect of sonar on whales and dolphins is related to both the frequency and intensity of the sound. Lower-frequency sounds travel farther through the water, while higher-frequency sounds drop off quickly. Furthermore, hearing sensitivity varies among species. Some species, such as baleen whales, tend to communicate and hear best in low frequencies (less than 1 kilohertz [kHz]), while others have the most sensitive hearing in mid or high frequencies. The most common sonars in the ocean are actually the fish finders and depth sounders on small vessels. These tend to use relatively high frequencies; common ones transmit at 50 kHz and 200 kHz, so the sounds do not travel very far. All odontocetes can probably hear the 50 kHz signals, and for some species it might be in their range of highest sensitivity, while 200 kHz sonars are likely above the hearing range of all species in Hawai'i.

The U.S. Navy also uses a wide diversity of types of sonar, but the ones of greatest concern to marine mammals are the mid-frequency active (MFA) sonars.

These are the loudest sonars being used, and they transmit between about 2 and 9 kHz, which is in the most sensitive hearing range for many species of toothed whales and dolphins. There are at least two different types of MFA sonars the navy uses, the loudest of which has a source level of about 236 decibels (dB)[1] and transmits at between 2 and 4 kHz. These loud mid-frequency sounds carry a long way underwater—they are probably audible to most species of odontocetes at ranges of up to 100 km. For comparison, humpback whale song source levels can be up to about 174 dB, a supertanker is about 190 dB, and an earthquake is about 210 dB.

In terms of any long-term effects of sound on Hawai'i's whales and dolphins, actual changes in population size, or distribution of a species, there is not much of a baseline to use for comparison. High-intensity MFA sonar has been utilized by the U.S. Navy in Hawai'i regularly for probably forty-five years or more, and the sonars currently in use have been around since the 1980s. So it is possible that effects on highly sensitive species may have happened long in the past. Abundance estimates for most species of whales or dolphins in Hawaiian waters are all fairly recent, and the ability to assess trends (whether populations are increasing, decreasing, or stable) is limited, to say the least.

Some who believe there are few or no effects argue that if whales or dolphins were being killed by naval sonar use, we would see many more dead animals on the beaches. But this argument ignores some very important factors that limit the likelihood of a dead or injured whale or dolphin being found in Hawai'i. First, there are numerous tiger sharks in Hawai'i, and opportunities to feed on a dead dolphin or whale are probably quickly seized. One humpback that live stranded on Maui was being consumed by several tiger sharks while the whale was still alive with its front end on the beach. I suspect that most whales or dolphins that die in Hawaiian waters are quickly consumed by scavenging sharks, so the

1. In water, decibels are measured 1 m away from the source (dB re 1 μPa @ 1m), and it is important to note that dB in water and in air are not measured on the same scale. An equivalent level in air is approximately 61.5 dB lower than in water, so 235 dB in water equates to 173.5 in air. Also, the dB scale is logarithmic, so a 10 dB increase is a tenfold increase in power, but the perceived loudness doubles. So, 236 dB is perceived as twice as loud as 226 dB.

A group of 150 to 200 melon-headed whales in shallow water at Hanalei Bay, Kaua'i, July 3, 2004. Navy mid-frequency active sonar was used nearby the night before the whales moved into the bay, as well as during the morning when the whales were in the bay. Photo by Dennis Fujimoto/*The Garden Island*.

likelihood of a carcass being around long enough to wash up on a beach would be low. For the main Hawaiian Islands insular false killer whales, one population for which we have a good idea of both abundance and adult mortality rates, the proportion of those that are predicted to die each year and are found either floating dead or on a beach ranges from about 2 to 4 percent—the remaining 96 to 98 percent of those that die are never found by humans.

It is not just scavenging sharks that make it unlikely a dead or injured whale or dolphin will be found in Hawai'i. Much of the coastline in Hawai'i is made up of inaccessible sea cliffs (e.g., much of northern Moloka'i and Kaua'i or southeastern Hawai'i Island) or largely uninhabited (e.g., Ni'ihau, Kaho'olawe). In other areas there are large fringing reefs (e.g., the southern shore of Moloka'i and parts of O'ahu and Kaua'i) that would make it difficult for an animal to actually find its way to a beach, dead or alive. And many species live in deep water and sink when they die and will not refloat, as the intensive pressure of deep water counteracts any buoyancy created by decomposition. These include Cuvier's and Blainville's beaked whales, the species most likely to be affected by naval sonar due to their diving behavior and response to sonar.

Although my speculation is informed by what we know about the distribution, movements, and abundances of many of the species of whales and dolphins throughout the main Hawaiian Islands, it is only possible to speculate about what effects the naval sonars may have had on the resident populations of marine mammals over time. Hawai'i is the home of the U.S. Navy's Pacific Fleet, with major bases of operation on O'ahu and Kaua'i. While the navy's operational area in Hawai'i covers some 608,000 square km (235,000 square miles) of ocean and naval activities take place throughout most of Hawaiian waters, its training and testing activities are concentrated off of Kaua'i. The Pacific Missile Range Facility on Kaua'i has a series of almost two hundred bottom-mounted hydrophones that cover a large area reaching more than halfway between Kaua'i and Ni'ihau and extending about 100 km north to south. These hydrophones allow the navy to acoustically monitor their own submarines and surface vessels on the range, and they conduct many of their training activities there. Thus, whale and dolphin populations off Kaua'i and Ni'ihau are probably more regularly and predictably exposed to naval sonar and other naval activities than in most other areas in Hawai'i. Naval sonar use does occur elsewhere in Hawai'i, but on a less

A dwarf sperm that live stranded at the mouth of the Kīlauea Stream, Kaua'i, August 27, 2009, the same morning that a U.S. Navy Submarine Commanders Course had started about 50 km to the northwest. The individual, an adult male, had a full stomach, and the necropsy showed it was in good condition with no obvious abnormalities. Photo by Kim Steutermann Rogers.

frequent and less predictable basis. Given the long history of regular use of MFA sonar off Kaua'i and Ni'ihau, I suspect that the mix of species we do or do not regularly see there today reflects in part this history of exposure to MFA sonar. We know there are resident populations of at least four species of odontocetes off those islands—spinner, bottlenose, and rough-toothed dolphins, as well as short-finned pilot whales—and all but newborn individuals in those populations have likely been repeatedly exposed to MFA sonar at varying levels throughout their lives. Whether such exposure has influenced survival rates or birthrates in these populations is unknown, but their persistence in the area suggests that individuals in these populations must tolerate MFA sonar use to some extent. As we build up photo-identification catalogs, we will be able to begin looking at survival and birthrates in these populations and compare them to other island-associated populations in Hawai'i that are exposed to MFA sonar much less often, such as those off Hawai'i Island.

Exposure to MFA sonar at very close ranges—say tens of meters—could cause permanent hearing damage, and for species that use sound to find food and communicate, permanent hearing damage may have major consequences for survival. Animals can suffer temporary hearing loss at much greater distances—think of going to rock concerts in your youth—and repeated temporary loss can lead to permanent hearing loss, so even "tolerating" MFA sonar use can lead to long-term negative consequences. As I mentioned in the account for Blainville's beaked whales, evidence from the Bahamas suggests that females regularly exposed to MFA sonar may have reduced reproductive rates, also potentially leading to long-term population changes.

Off Kaua'i and Ni'ihau, we have documented resident populations of only four species of odontocetes. Off O'ahu and around Maui Nui, we know of resident populations of six species. Off Hawai'i Island there are resident populations of ten species of odontocetes. Why the difference among islands? While there are oceanographic differences among the island areas in Hawai'i, I suspect that the relative paucity of resident species off Kaua'i and Ni'ihau may be due, at least in part, to the higher levels of naval MFA sonar use off those islands and the long-term displacement of particularly sensitive species. Five of the six species that appear to be in very low density off Kaua'i and Ni'ihau and for which we have no evidence of resident populations versus those found off Hawai'i Island include those thought to be most sensitive to effects from MFA sonar. These are Cuvier's and Blainville's beaked whales, as well as others that are also thought to be sensitive to MFA sonar: dwarf sperm whales, pygmy killer whales, and melon-headed whales. Pantropical spotted dolphins are also seen only infrequently off Kaua'i and Ni'ihau, while being one of the most frequently encountered species off the other islands. No one knows whether they are sensitive to MFA sonar; finding out, through controlled playbacks of simulated sonar sounds as is being done for many other species of whales and dolphins in areas such as southern California, would be an important research goal. Given that resident populations of these species may have been displaced long ago, it is not likely possible to prove whether regular MFA sonar exposure over the last thirty-plus years is actually responsible for these differences in the presence and abundance of resident species.[2] But these observations underscore the need for cautious and careful management of MFA sonar impacts, both in areas where it is regularly used and in areas where it is only infrequently used.

There is no doubt at this stage that some species of whales and dolphins are much more likely to be affected by MFA sonar than others. Even within a species, there can be a substantial difference in the reactions of a particular individual or group of animals to MFA sonar: some may react quite strongly, and others may show little or no change in behavior. This likely varies not only by how close they are when it is first turned on but how novel the exposure is, what the individuals are doing at the time of exposure, and what the particular circumstances are of the exposure. For example, was the vessel using MFA sonar heading toward or away from the animals? For species that feed on patchy prey, if there is a high concentration of prey in an area, it's possible that animals exposed to MFA sonar will be less likely to leave the area than if there is a lower concentration of prey and thus potentially place themselves at greater risk of temporary hearing damage. As I mentioned in the species accounts for Cuvier's and Blainville's beaked whales, behavioral reactions to MFA sonar can occur at great distances from the

2. There are resident populations of Cuvier's beaked whales off southern California and Blainville's beaked whales in the Bahamas that are also regularly exposed to MFA sonar and have not shown signs of abandonment, but those areas have much higher densities of both species of whale, suggesting they may be particularly good foraging sites and thus may be less likely to be abandoned.

source, and these types of reactions, repeated over and over again, can have long-term implications. Naïve animals may be more at risk to the types of behavioral reactions that could lead to decompression-like sickness or potentially placing themselves in a situation where the likelihood of stranding is higher.

What can be done about this? The U.S. Navy is mandated to minimize or mitigate their impacts on whales and dolphins, when they can do so without compromising their mission. Much of their current mitigation is composed of posting a watch to inform the bridge crew on navy vessels if marine mammals are sighted within a certain range and turning off MFA sonar if such sightings occur. Given that most species of whales and dolphins likely react to MFA sonar well before they are within visual range of an observer, and that even animals close to a navy vessel could be missed by an observer in the very rough seas typically experienced during navy exercises in Hawai'i, I think these watch standers are an ineffective way to mitigate effects of MFA sonar.

Our research has found relatively few resident populations of dolphins and whales in areas of high MFA sonar use and more than twice the number of resident populations in an area with much less sonar use. This leads to one obvious mitigation measure: sonar exclusion areas (or time-area restrictions). A court decision from recent lawsuits brought by the Natural Resources Defense Council (NRDC) and Conservation Council for Hawai'i (CCH) against NMFS and the U.S. Navy over sonar use off Hawai'i and southern California specifically noted the lack of exclusion areas there. Our work, particularly in determining home ranges of populations and designating "Biologically Important Areas," has played a role in these lawsuits. There has been tremendous resistance within the U.S. Navy and NMFS to set aside areas where MFA sonar is excluded, yet that seems likely the only effective way of protecting populations of species that are particularly sensitive to impacts. These include the small resident populations of Cuvier's beaked whales, Blainville's beaked whales, and melon-headed whales off Hawai'i Island. The judge noted that the U.S. Army does not need "to continuously occupy every square mile of land within the United States," so the navy "cannot possibly need to do so" either. In September 2015 a settlement for these lawsuits was announced between NRDC, CCH, NMFS, and the U.S. Navy that provides for some restrictions of MFA sonar and explosive use around Hawai'i Island as well as off Maui Nui. While claimed as a "victory" by environmental groups, I'm not convinced these restrictions will greatly reduce the exposure of the resident whales and dolphins to MFA sonar use off the island, and thus it may not provide any substantive protection for them.

WATCHING DOLPHINS AND WHALES

Some other human interactions with marine mammals are more deliberate. In the case of dolphin and whale watching or swimming with dolphins, the intent is not to cause harm, but like most things, there are good ways and bad ways to go about it. There are over 8 million visitors to Hawai'i each year, and an estimated 22 to 31 percent of them—somewhere between 1.7 and 2.5 million people—end up on boating tours or whale-watching tours. While many of those are not specifically directed toward wildlife, many boats out on the water, including sport fishing charters, will divert from their path to watch whales or dolphins. The likelihood of accidentally harassing dolphins or whales varies by species (some like boats, some don't), as well as the type of interaction (how close the vessel or swimmer gets, and how they behave toward the animals). There is no doubt that humans getting close to dolphins or whales can result in harassment, and repeated instances of harassment on small populations can have detrimental consequences; I noted some of these potential consequences earlier in the spinner dolphin section. With the exception of humpback whales, where there is an approach regulation in Hawai'i (deliberately approaching within 100 yards is illegal), for all other species there are only guidelines for how close to approach. Harassment itself is illegal, but proving harassment in any particular case can be difficult, and there is virtually no enforcement of the harassment regulation in Hawai'i. In the case of dolphins and small whales, the NMFS guidelines suggest that humans and vessels should remain 50 yards away, while for large whales it is 100 yards.

While simple, a one-size-fits-all guideline is not an effective way to keep swimmers or boaters from harassing animals, and it does not make sense to many of the people involved in the interactions, particularly when some species and some subsets of individuals regularly approach boats or swimmers to interact with them. Among the Hawaiian blackfish, for example, false killer whales often approach boats to bowride or interact with them, melon-headed whales

Spinner dolphins and a free diver off Kona, Hawai'i. Photo by Bo Pardau.

occasionally do so (particularly juveniles and subadults), and pilot whales rarely do. Pygmy killer whales usually avoid boats to some degree, although there are always exceptions. So advising people to stay 50 yards away from pilot whales or pygmy killer whales makes sense to me, but doing so with false killer whales is probably neither effective nor necessary, as the animals usually seek out boats and are often curious about people in the water. I don't mean to say that human interactions with species like false killer whales *can't* disrupt important behaviors such as feeding, resting, or nursing; they can, but guidelines should take into account that some species will regularly approach boats or swimmers. Many species of dolphins regularly bowride, including bottlenose, spinner, and spotted dolphins, the three most frequently encountered in Hawai'i. Often, however, it is just a subset of individuals within a group, particularly juveniles, that do so, giving people the impression that *all* the animals want to interact with them. This is the root of some of the conflicts that arise with many swim-with-dolphin tour companies or individuals who go out from shore to swim with spinner dolphins. Individuals in the water may see one or two dolphins approaching out of curiosity but miss the fact that the main body of the group of dolphins may have their sleep disrupted and are repeatedly having to move away from swimmers.

But I think there is—or can be—a strong benefit to having people experience whales or dolphins close up in the wild. Such interaction with wild animals, or simply observations of them, can have a lasting impact on how people view wildlife and how much they are willing to change their personal behavior in ways that could benefit whales, dolphins, and other wildlife (for example, limiting indiscriminate use of pesticides or herbicides). Connections with nature, whether spiritual or scientific, are key to having a constituency of people willing to voice their opinions, cast their votes, and put their money or time toward conservation of wildlife and wild ecosystems. I think it is vitally important that people have the opportunity to experience whales and dolphins in the wild in Hawai'i, but—and this is critical—it has to be in a respectful way. Tour companies taking people out to see or interact with animals should include a message that both educates people and inspires them to want to conserve and protect the animals they are seeing.

But how does this translate in real life? It is unlikely that NMFS is going to regulate companies so that they must include a rigorous conservation message on their dolphin- or whale-watching or swim-with-dolphin trips. I think there is a lot of value in working with tour operators and naturalists, and working with members of the public who regularly interact with wildlife, to educate them about the best way to approach different species and to recognize behaviors that may indicate an animal is being disturbed. There was an effort in Kona a number of years ago, spearheaded by the Coral Reef Alliance, to develop community guidelines on ways of interacting with marine wildlife, including dolphins and whales. That effort was put on hold as NMFS was expected to put out new rules for interacting with spinners, but the approach was a good one, as it involved working with all the operators to develop voluntary standards and guidelines. Voluntary standards alone, however, are not enough, and they need to be combined with other approaches, including enforceable regulations.

Up to the present, the NMFS approach to the swim-with-dolphin businesses in Hawai'i has basically been to discourage them. But this approach has not been particularly effective: as of early 2016 there are about twenty-eight tour companies off Kona that take people out to swim with dolphins, as well as about ten off O'ahu, two off Maui, and one off Kaua'i. There are also businesses that offer spiritual retreats that include dolphin swims, with about nineteen on Hawai'i Island, one on O'ahu, seven on Maui, and at least two on Kaua'i. Unless it is possible to prohibit swimming with all species of dolphins in all areas at all times (making it a regulation rather than a guideline), I believe a more productive approach would be to work with the many companies in Hawai'i that offer such programs to help them do it in a way that minimizes harassment. Just saying it is "bad" does not convince people to stop doing it.

In 2005, NMFS published an "advance notice of proposed rulemaking" regarding potentially implementing regulations to protect spinner dolphins from harassment. However, as of early 2016 no proposed rules had been released, although there is talk about a proposed rule coming out in 2016. When they first released the "advance notice," one unintended consequence was encouraging some vessel operators that were focusing on spinner dolphins to look farther offshore for interactions with other species of dolphins or whales. I hope that if or when new regulations are put in place, such potential for shifting the problem to other species is taken into account.

CONSERVATION OF FALSE KILLER WHALES IN HAWAI'I

Although the process of conservation (taking some action once the threats or problems have been identified) seems straightforward, typically nothing happens quickly in terms of actually implementing conservation or management measures. And nothing has brought this home as strongly as the case of false killer whales in Hawai'i. Starting in the 1960s, at least a dozen false killer whales were captured in Hawai'i and taken into captivity for aquariums. In 1989, several aquariums were interested in capturing false killer whales again, but capture permits were harder to get, and there was not enough known about how many there were in Hawai'i to justify a permit. That year biologists Steve Leatherwood and Randy Reeves flew aerial surveys off O'ahu, Lāna'i, and Hawai'i Island to try to get a minimum count of false killer whales around the islands. Steve and Randy saw seven different species of odontocetes in eight days of flying in June and July 1989, and on three different days they observed groups of over 300 false killer whales, with the largest group of 470. They wrote a short report on their surveys, but it was never published, and while a capture permit was issued, no more animals were captured. With the lack of interest in false killer whales, or for that matter any Hawaiian odontocete other than spinner dolphins at that point, the report dropped off the radar.

By 1999, observer data from the Hawai'i-based longline fishery was showing that bycatch of false killer whales was higher than any other whale or dolphin species. However, at that stage there was no abundance estimate with which to compare these bycatch data. In 2000, Joe Mobley from the University of Hawai'i produced an abundance estimate for false killer whales of only about 121 individuals from aerial surveys flown in the 1990s around the main Hawaiian Islands. When NMFS prepared the stock assessment report for false killer whales in 2000, they were considered a "strategic stock," with bycatch (from the longline fishery) that was not sustainable. Such high levels of bycatch should have resulted in the longline fishery being recategorized from one that had a "remote likelihood of or no known" incidental mortality to one that had "frequent" incidental mortality. This would have allowed for the formation of a Take Reduction Team and production of a Take Reduction Plan to reduce bycatch. It was not until 2004 that the fishery was recategorized, and only after a lawsuit brought by Earthjustice on behalf of a local Hawaiian group, Hui Mālama i Kohalā (which translates loosely as "a group of people that cares for whales"), along with the Center for Biological Diversity.

By 2003, the first genetic results were available, based on the biopsy sampling we had started in 2000. These results revealed that false killer whales around the main Hawaiian Islands are genetically differentiated from false killer whales elsewhere in the Pacific. In 2005, we produced our first abundance estimate based on photo identification, and it was very similar to Joe Mobley's aerial survey estimate, with just 123 individual false killer whales around the main Hawaiian Islands. It was now clear that the Hawai'i whales were part of a small population that did not interbreed with false killer whales elsewhere. Each subsequent year, observer data from the longline fishery continued to show that false killer whales were the species of whale or dolphin killed or seriously injured most often in the fishery, and that bycatch continued to exceed the sustainability threshold, but there was still no progress on formation of a Take Reduction Team.

The recategorization of the longline fishery in 2004, combined with the 2005 stock assessment report showing that high levels of bycatch in the fishery continued, should have triggered formation of a Take Reduction Team. Although progress was made on the science front in intervening years, it was not until another lawsuit brought by Earthjustice in 2009 that NMFS finally announced, in January 2010, that a Take Reduction Team was being formed. The team included longline fishermen, scientists, conservationists, and representatives of the State of Hawai'i government. We met four times between February and July 2010 and came to a consensus on a draft Take Reduction Plan, which was submitted to NMFS. As the management agency in charge, it was up to NMFS to finalize the plan and put it into effect. As I said, nothing happens quickly. It was not until December 2012 that the first parts of the plan were put in place, with some of the measures to reduce bycatch not taking effect until the end of February 2013.

Steve Leatherwood passed away in 1997, a couple years before I started working in Hawai'i. I had seen the report he and Randy Reeves had written, but it was not until 2007 that its importance became clear to me: it provided early evidence of a population decline. Throughout the early 2000s, evidence continued to come in from the longline observers of high levels of bycatch, and our photo-identification work was showing individuals from the main Hawaiian Islands population with

A false killer whale with an injured dorsal fin, likely a line injury from a fishery interaction, off O'ahu, May 17, 2014. This individual, HIPc186 in our photo-identification catalog, is a member of Cluster 3 of the endangered main Hawaiian Islands population. Photo by Chuck Babbitt/babbittphoto.com.

line injuries on the dorsal fins that we attributed to fisheries interactions. The 1989 report was largely forgotten in part because it was not published and it was written at a time when there was nothing known about false killer whales in Hawai'i (or, for that matter, almost anywhere). In January 2008, I contacted Randy about writing up the results for publication in a peer-reviewed journal. Peer-reviewed science is the "gold standard" and once something is published in a peer-reviewed journal it is hard for management agencies such as NMFS to ignore. Peer-reviewed papers also carry a lot of weight in the courts, and often lawsuits and court decisions seem to drive what actions management agencies eventually take. In early 2008, I worked with Randy to write up the results, and the paper, "Evidence of a Possible Decline since 1989 in False Killer Whales (*Pseudorca crassidens*) around the Main Hawaiian Islands," was published in 2009.

Other results from our work were also pointing toward a conservation problem with false killer whales in Hawaiian waters. Working with Gina Ylitalo at the Northwest Fisheries Science Center, analyses of biopsy samples were revealing that all the adult male false killer whales we had sampled had PCB levels that were exceeding a health threshold. These levels could be causing immune suppression, which would increase the mortality rates of males. Reanalyzing Joe Mobley's aerial survey data from the 1990s and early 2000s showed a statistically significant decline in false killer whale sighting rates over a ten-year period. Based on all of these findings, NMFS could have initiated a review of the status of false killer whales in Hawai'i to determine whether they should be listed under the Endangered Species Act. Such reviews are rarely initiated by NMFS without external prodding, and in September 2009 NRDC submitted a petition to NMFS to list the insular population of false killer whales under the ESA. In January 2010, NMFS announced they were going to conduct a status review, and that status review was published in August of the same year. The August 2010 status review did recognize the main Hawaiian Islands insular population as a "Distinct Population Segment" and concluded that it was "significant" in comparison to the global population and at risk of extinction, but it was more than two years before NMFS announced a final listing decision. That decision was announced only after a lawsuit was filed by NRDC. In November 2012, NMFS announced that the insular population would be listed as endangered under the ESA. As of early 2016, however, this listing had had few implications for the population, although NMFS has plans to set up a Recovery Team and put together a Recovery Plan, as well as designate critical habitat.

The Take Reduction Plan includes one measure that should provide increased protection for the main Hawaiian Islands insular population of false killer whales. The longline exclusion zone around the main Hawaiian Islands formerly was closer to the islands four months of the year, with some of it coming as close as 45 km (24 nautical miles [nm]) from shore. However, since 2013 that boundary has been fixed in place, so the closest longline fishing permitted is at least 78 km (42 nm) from shore. From our satellite tagging work, it appears that the main

Hawaiian Islands false killer whales only rarely overlap with areas where longline fishing occurs.

Other measures put into place as part of the Take Reduction Plan include using "weak" circle hooks and strong terminal gear. When a fish tries to take the bait off a circle hook, they are often hooked in the corner of the mouth, rather than ingesting the hook. When false killer whales attempt to take fish off of hooks, they often leave the head attached, so if a fish is hooked in the jaw the chances of the false killer whale being hooked are lower, assuming it leaves the head of the fish on the hook. With strong terminal gear (the leaders and lines attached to the hooks) and weaker hooks, the hooks are also more likely to straighten out and allow the whale to get free if hooked in the mouth without trailing a lot of fishing gear behind it. Both of these factors should reduce the injury and mortality rate of false killer whales interacting with longline gear. Whether these measures are working, however, is not clear. The hooking rates of pelagic false killer whales in the longline fishery have remained high, at least through the end of 2015.

LESSONS LEARNED

These conservation issues—the impacts of naval sonar on many species and the specific threats faced by false killer whales and how to mitigate them—have many things in common. Science and monitoring, most of it ongoing, is what led to the identification of the issues and determined that the populations were particularly vulnerable. Environmental groups suing or petitioning the federal government, whether it is NMFS or the U.S. Navy, have played a major role in conservation and management for both of these issues. Peer-reviewed publications from the scientific community have played a big role in both cases,

‹‹‹

A dead adult male false killer whale found stranded at Ka Lae, Hawai'i Island, October 5, 2013. This individual, HIPc162 in our photo-identification catalog, was a member of Cluster 3 of the main Hawaiian Islands population. A necropsy was undertaken by Hawai'i Pacific University, although cause of death was not determined. A broken tooth is visible, and this individual had five fishhooks in the stomach. Photo by Tom Elliot.

A spinner dolphin found entangled in a net fragment off Kona, March 10, 2014. Photo by Deron S. Verbeck/iamaquatic.com.

both in terms of convincing the courts that there is rigorous science behind the environmental concerns and providing a strong, defensible basis for the management agency (NMFS) to take action. In both cases, as I have stated before, nothing has happened quickly, and I have left out many of the steps involved in both processes. There has been a considerable amount of education required in these cases, within and among the scientific community, environmental groups, fishermen, and government agencies, as well as the general public. Working with all of these groups to better understand the issues faced by whale and dolphin populations in Hawai'i is critical to helping conserve and protect them. This is certainly going to continue to be the case with other issues, such as harassment associated with swimming with spinner dolphins or other species, as well as mitigating the effects of fisheries interactions.

Whale and dolphin interactions with fisheries in Hawai'i take a variety of forms. These include animals coming across discarded net fragments and becoming entangled in them; species seeking out fishing vessels to take their bait or catch (see the sections on rough-toothed dolphins and false killer whales) and occasionally getting hooked as a result; some small subset of fishermen occasionally shooting at animals in retaliation or to discourage them from taking their bait or catch; and fishermen seeking out dolphins, particularly pantropical spotted dolphins, to catch associated tuna. Understanding and potentially mitigating negative interactions with fishermen are complicated by the fact that there are over three thousand individuals in Hawai'i with "commercial marine licenses"—effectively, these are commercial fishing licenses. These licenses cost only $50 a year for residents, and many of these "commercial" fishermen are simply recreational fishermen who occasionally hit the jackpot, finding floating debris and catching more *mahi mahi* than will fit in their freezers: these licenses allow for selling catch to restaurants or buyers. Some of these license holders are captains of charter vessels that take people—primarily tourists—out to catch game fish. Others are full-time commercial or subsistence fishermen using a variety of types of gear and fishing methods. Some, like the charter vessels, most or all of the recreational fishermen, and a lot of the commercial fishermen, fish only during the day, while many commercial fishermen fish at night as well. There are dozens of different methods of fishing, and these vary among islands and even off different sides of the islands. There is no one "fishing community" in Hawai'i. The ethnic makeup of fishermen is extremely diverse, as are the motivations to fish and the attitudes toward conservation, whether it be sustaining fish populations for their children and grandchildren or acceptance of the coexistence of other species that also subsist on the same fish populations, such as false killer whales. While science, both environmental and social, is critical for understanding the implications of the conflicts between fishermen and whales and dolphins in Hawaiian waters, mitigating those conflicts will require a long-term approach. Working with fishermen to find solutions will be the key, and educating and inspiring new generations of fishermen to accept the role of dolphins and whales in the ocean ecosystem is essential.

Fishing vessels regularly troll through groups of pantropical spotted dolphins trying to catch tuna that may be associated with the dolphins, and dolphins are occasionally hooked. This photo was taken off Kona, May 11, 2011. Photo by Annie B. Douglas.

CONSERVATION

Despite progress in recent years, there remain many challenges to conserving and managing whale and dolphin populations in Hawaiian waters. I am often asked what people can do to help, and it is clear that individuals—you, reading this book—can and need to play a role. Recognize that Hawaiian waters are an oasis for many populations of whales and dolphins, as well as other marine life, and that the oasis needs protection. Eat sustainably caught local fish, species that grow quickly and are fast to reproduce, such as *mahi mahi* and *ono,* and make sure they are caught with methods that reduce bycatch of other species. Educate yourself about fisheries, pollutants, and the impacts of sound, and once you have done that, educate your friends and family. Make sure you vote, but before that contact your elected representatives about protecting the marine environment, and make it clear to them you won't vote for a politician who doesn't work hard to protect Hawai'i's marine wildlife. Keep plastics and bits of nets, as well as pesticides and herbicides, out of the water, and minimize their use. If you are fishing yourself, use gear that is less likely to hook nontarget species (circle hooks?), and recognize that most species of whales and dolphins in Hawaiian waters are *kama'āina*—native-born residents of Hawai'i—and were here long before the Polynesians first discovered the islands.

ACKNOWLEDGMENTS

This book would not have been possible without the help of a lot of people. In the field, I need to especially thank Daniel Webster, who has played a critical role over the last sixteen years, as well as Jess Aschettino, Annie Douglas, Annie Gorgone, Dan McSweeney, Brenda Rone, Greg Schorr, Allan Ligon, Russ Andrews, and a large number of others who have participated in our work. In the office, Sabre Mahaffy has also played a critical role curating most of our photo-ID catalogs and examining social organization of most species, with the help of many others including Annie Gorgone, Jess Aschettino, and numerous interns from Cascadia Research. Much of what we know about many species comes from contributions by a number of collaborators, particularly Dan McSweeney and Tori Cullins, but with special mention to the late Dan Salden, Deron Verbeck, Mark Deakos, Chris Bane, and Chuck Babbitt. I also want to thank a large and growing number of individuals who spend time on the water and are contributing photos: they play a very important role in understanding the resident populations in Hawaiian waters. One of the reasons our work has been so productive is the large number of collaborators working with data, recordings, or samples: Melanie Abecassis, Renee Albertson, Russ Andrews, Jay Barlow, Simone Baumann-Pickering, Amanda Bradford, Susan Chivers, Sarah Courbis, Kerry Foltz, Brad Hanson, Brenda Jensen, Max Kaplan, Karen Martien, Aran Mooney, Erin Oleson, Jeff Polovina, Amy Van Cise, Bill Walker, Kristi West and Gina Ylitalo, to name most of them.

ACKNOWLEDGMENTS

I want to thank Amanda Bradford, Karen Martien, Bruce Mate, and Kristi West for providing unpublished information, Greg Donovan and the International Whaling Commission for providing locations of whaling records, as well as a number of other individuals for answering questions as the writing progressed or providing photos: Melanie Abecassis, Kristen Ampela, Annalisa Berta, Bob Brownell, Phil Clapham, Diane Claridge, Dennis Fujimoto, Marie Hill, Tom Jefferson, John Kelly, Adam Kurtz, Matt Leslie, Gretchen Lovewell, Ed Lyman, David Mann, Laura McCue, Bill McLellan, Aliza Milette-Winfree, Sally Mizroch, Erin Oleson, Kara Osada, Ann Pabst/UNCW Marine Mammal Stranding Program, Bill Perrin, Bob Pitman, Karen Pryor, Shannon Rankin, Susan Rickards, Ed Shallenberger, Fred Sharpe, Ana Širovi, Julie Steelman, Trisha Kehaulani Watson-Sproat, Randy Wells, Carlie Wiener, Bernd Würsig, Suzanne Yin, and Ann Zoidis. Dave Janiger's reference database and willingness to track down obscure references has been extremely helpful for this effort, and I thank Else Jean Jensen for providing an English translation of an older Danish publication. Thanks are due to the many individuals who have come out on the water with us over the years, both providing help in the field and asking many questions that I've tried to answer in this book.

In terms of reviews, I want to thank the two external reviewers, Bob Brownell and Randy Wells, as well as Laurie Shuster, for reviewing all of this book, as well as Dave Anderson for reviewing most of it, and a number of individuals for reviewing certain sections: Hal Whitehead (sperm whales), Renee Albertson (rough-toothed dolphins), Jessica Aschettino (melon-headed whales), Michael Jasny (conservation), Sabre Mahaffy (sperm whales and beaked whales), Susan Rickards (spinner dolphins and conservation), Amy Van Cise (pilot whales), and Fred Sharpe (sperm whales and baleen whales). I also want to especially thank Annie Douglas for reviewing the entire manuscript several times, for sorting through and pulling out photographs, both for inclusion and for assessing color patterns and wounds, and for providing feedback throughout the writing process. Uko Gorter deserves thanks for his illustrations and his willingness to put up with multiple rounds of suggestions for making them fit Hawaiian populations better. Similarly, I thank Dave Anderson for going through numerous iterations of maps to produce many of those in this book.

My research described in this book was funded over the years by a variety of sources. The long-term ones have been the National Marine Fisheries Service (particularly the Southwest Fisheries Science Center, the Pacific Islands Fisheries Science

Center, and the Northwest Fisheries Science Center), the U.S. Navy (Living Marine Resources, Pacific Fleet, and Office of Naval Research), and Dolphin Quest. Funding or support has also been received from the Hawai'i Ocean Project, the Cascadia Research Collective, the Hawaiian Islands Humpback Whale National Marine Sanctuary, the John F. Long Foundation, the Marine Mammal Commission, the M. B. & Evelyn Hudson Foundation, and the Wild Whale Research Foundation. The research described was undertaken under a number of different NMFS Scientific Research Permits (Nos. 926, 731-1509, 731-1774, 774-1714, 782-1719, 1036-1699, 14097, 15330, and 16163). Photos of stranded animals were taken under the MMHSRP Permit No. 932-1489 or 932-1905.

SOURCES AND FURTHER READING

Author's note: For further information, please access the Cascadia Research Collective's Web site: www.cascadiaresearch.org/hawaii.htm.

Albertson, G. R. 2014. "Worldwide Phylogeography and Local Population Structure of the Rough-Toothed Dolphin (*Steno bredanensis*)." PhD thesis, Oregon State University.

Andrews, K. R., L. Karczmarski, W. W. L. Au, S. H. Rickards, C. A. Vanderlip, B. W. Bowen, E. G. Grau, and R. J. Toonen. 2010. "Rolling Stones and Stable Homes: Social Structure, Habitat Diversity and Population Genetics of the Hawaiian Spinner Dolphin (*Stenella longirostris*)." *Molecular Ecology* 19:732–748.

Aschettino, J. M., R. W. Baird, D. J. McSweeney, D. L. Webster, G. S. Schorr, J. L. Huggins, K. K. Martien, S. D. Mahaffy, and K. L. West. 2011. "Population Structure of Melon-Headed Whales (*Peponocephala electra*) in the Hawaiian Archipelago: Evidence of Multiple Populations Based on Photo-Identification." *Marine Mammal Science* doi: 10.1111/j.1748-7692.2011.00517.x.

Bachman, M. J., J. M. Keller, K. L. West, and B. A. Jensen. 2014. "Persistent Organic Pollutant Concentrations in Blubber of 16 Species of Cetaceans Stranded in the Pacific Islands from 1997 through 2011." *Science of the Total Environment* 488–489:115–123.

Baird, R. W. 2005. "Sightings of Dwarf (*Kogia sima*) and Pygmy (*K. breviceps*) Sperm Whales from the main Hawaiian Islands." *Pacific Science* 59:461–466.

Baird, R. W., D. Cholewiak, D. L. Webster, G. S. Schorr, S. D. Mahaffy, C. Curtice, J. Harrison, and S. M. Van Parijs. 2015. "Biologically Important Areas for Cetaceans within U.S. Waters—Hawai'i Region." *Aquatic Mammals* 41:54–64.

Baird, R. W., A. M. Gorgone, D. J. McSweeney, A. D. Ligon, M. H. Deakos, D. L. Webster, G. S. Schorr, K. K. Martien, D. R. Salden, and S. D. Mahaffy. 2009. "Population Structure of Island-Associated Dolphins: Evidence from Photo-Identification of Common Bottlenose

Dolphins (*Tursiops truncatus*) in the Main Hawaiian Islands." *Marine Mammal Science* 25:251–274.

Baird, R. W., A. M. Gorgone, D. J. McSweeney, D. L. Webster, D. R. Salden, M. H. Deakos, A. D. Ligon, G. S. Schorr, J. Barlow, and S. D. Mahaffy. 2008. "False Killer Whales (*Pseudorca crassidens*) around the Main Hawaiian Islands: Long-Term Site Fidelity, Inter-Island Movements, and Association Patterns." *Marine Mammal Science* 24:591–612.

Baird, R. W., M. B. Hanson, G. S. Schorr, D. L. Webster, D. J. McSweeney, A. M. Gorgone, S. D. Mahaffy, D. Holzer, E. M. Oleson, and R. D. Andrews. 2012. "Assessment of Range and Primary Habitats of Hawaiian Insular False Killer Whales: A Scientific Basis for Determination of 'Critical Habitat.'" *Endangered Species Research* 18:47–61.

Baird, R. W., A. D. Ligon, S. K. Hooker, and A. M. Gorgone. 2001. "Subsurface and Night-time Behaviour of Pantropical Spotted Dolphins in Hawai'i." *Canadian Journal of Zoology* 79:988–996.

Baird, R. W., S. D. Mahaffy, A. M. Gorgone, T. Cullins, D. J. McSweeney, E. M. Oleson, A. L. Bradford, J. Barlow, and D. L. Webster. 2014. "False Killer Whales and Fisheries Interactions in Hawaiian Waters: Evidence for Sex Bias and Variation among Populations and Social Groups." *Marine Mammal Science* doi: 10.1111/mms.12177.x.

Baird, R. W., D. J. McSweeney, C. Bane, J. Barlow, D. R. Salden, L. K. Antoine, R. G. LeDuc, and D. L. Webster. 2006. "Killer Whales in Hawaiian Waters: Information on Population Identity and Feeding Habits." *Pacific Science* 60:523–530.

Baird, R. W., E. M. Oleson, J. Barlow, A. D. Ligon, A. M. Gorgone, and S. D. Mahaffy. 2013. "Evidence of an Island-Associated Population of False Killer Whales (*Pseudorca crassidens*) in the Northwestern Hawaiian Islands." *Pacific Science* 67:513–521.

Baird, R. W., G. S. Schorr, D. L. Webster, S. D. Mahaffy, D. J. McSweeney, M. B. Hanson, and R. D. Andrews. 2011. "Open-Ocean Movements of a Satellite-Tagged Blainville's Beaked Whale (*Mesoplodon densirostris*): Evidence for an Offshore Population in Hawai'i?" *Aquatic Mammals* 37:506–511.

Baird, R. W., G. S. Schorr, D. L. Webster, D. J. McSweeney, M. B. Hanson, and R. D. Andrews. 2011. "Movements of Two Satellite-Tagged Pygmy Killer Whales (*Feresa attenuata*) off the Island of Hawai'i." *Marine Mammal Science* doi: 10.1111/j.1748-7692.2010.00458.x.

Baird, R. W., D. L. Webster, J. M. Aschettino, G. S. Schorr, and D. J. McSweeney. 2013. "Odontocete Cetaceans around the Main Hawaiian Islands: Habitat Use and Relative Abundance from Small-Boat Sighting Surveys." *Aquatic Mammals* 39:253–269.

Baird, R. W., D. L. Webster, S. D. Mahaffy, D. J. McSweeney, G. S. Schorr, and A. D. Ligon. 2008. "Site Fidelity and Association Patterns in a Deep-Water Dolphin: Rough-Toothed Dolphins (*Steno bredanensis*) in the Hawaiian Archipelago." *Marine Mammal Science* 24:535–553.

Baird, R. W., D. L. Webster, G. S. Schorr, D. J. McSweeney, and J. Barlow. 2008. "Diel Variation in Beaked Whale Diving Behavior." *Marine Mammal Science* 24:630–642.

Balcomb, K. C., and D. E. Claridge. 2001. "A Mass Stranding of Cetaceans Caused by Naval Sonar in the Bahamas." *Bahamas Journal of Science* 01/05:2–12.

Barlow, J. 2006. "Cetacean Abundance in Hawaiian Waters Estimated from a Summer/Fall Survey in 2002." *Marine Mammal Science* 22:446–464.

Barlow, J., J. Calambokidis, E. A. Falcone, C. S. Baker, A. M. Burdin, P. J. Clapham, J. K. B. Ford,

C. M. Gabriele, R. LeDuc, D. K. Mattila, T. J. Quinn, L. Rojas-Bracho, J. M. Straley, B. L. Taylor, J. Urbán R., P. Wade, D.Weller, B. H. Witteveen, and M. Yamaguchi. 2011. "Humpback Whale Abundance in the North Pacific Estimated by Photographic Capture-Recapture with Bias Correction from Simulation Studies." *Marine Mammal Science* doi: 10.1111/j.1748-7692.2010.004444.x.

Baumann-Pickering, S., M. A. McDonald, A. E. Simonis, A. S. Berga, K. P. B. Merkens, E. M. Oleson, M. A. Roch, S. M. Wiggins, S. Rankin, T. M. Yack, and J. A. Hildebrand. 2013. "Species-Specific Beaked Whale Echolocation Signals." *Journal of the Acoustical Society of America* 134:2293–2301.

Bradford, A. L., and K. A. Forney. 2014. "Injury Determinations for Cetaceans Observed Interacting with Hawaii and American Samoa Longline Fisheries during 2007–2011." NOAA Technical Memorandum NMFS-PIFSC-39.

Bradford, A. L., K. A. Forney, E. M. Oleson, and J. Barlow. 2014. "Accounting for Subgroup Structure in Line-Transect Abundance Estimates of False Killer Whales (*Pseudorca crassidens*) in Hawaiian Waters." *PLOS ONE* 9:e90464.

———. Forthcoming. "Line-transect Abundance Estimates of Cetaceans in the Hawaiian EEZ."

Brownell, R. L., Jr., C.-J. Yao, C.-S Lee, and M.-C. Wang. 2009. "Worldwide Review of Pygmy Killer Whales, *Feresa attenuata,* Mass Strandings Reveals Taiwan Hot Spot." International Whaling Commission Document SC/61/SM1.

Carretta, J. V., E. M. Oleson, D. W. Weller, A. R. Lang, K. A. Forney, J. Baker, B. Hanson, K. Martien, M. M. Muto, T. Orr, H. Huber, M. S. Lowry, J. Barlow, D. Lynch, L. Carswell, R. L. Brownell Jr., and D. K. Mattila. 2014. "U.S. Pacific Marine Mammal Stock Assessments, 2013." NOAA Technical Memorandum NMFS-SWFSC-532.

Claridge, D. E. 2013. "Population Ecology of Blainville's Beaked Whales (*Mesoplodon densirostris*)." PhD thesis, University of St. Andrews.

Courbis, S., R. W. Baird, F. Cipriano, and D. Duffield. 2014. "Multiple Populations of Pantropical Spotted Dolphins in Hawaiian Waters." *Journal of Heredity* 105:627–641.

Cox, T. M., T. J. Ragen, A. J. Read, E. Vos, R. W. Baird, K. Balcomb, J. Barlow, J. Caldwell, T. Cranford, L. Crum, A. D'Amico, G. D'Spain, A. Fernández, J. Finneran, R. Gentry, W. Gerth, F. Gulland, J. Hildebrand, D. Houser, T. Hullar, P. D. Jepson, D. Ketten, C. D. MacLeod, P. Miller, S. Moore, D. Mountain, D. Palka, P. Ponganis, S. Rommel, T. Rowles, B. Taylor, P. Tyack, D. Wartzok, R. Gisiner, J. Mead, and L. Benner. 2006. "Understanding the Impacts of Anthropogenic Sound on Beaked Whales." *Journal of Cetacean Research and Management* 7:177–187.

D'Amico, A., and R. Pittenger. 2009. "A Brief History of Active Sonar." *Aquatic Mammals* 35:426–434.

DeRuiter, S. L., B. L. Southall, J Calambokidis, W. M. X. Zimmer, D. Sadykova, E. A. Falcone, A. S. Friedlaender, J. E. Joseph, D. Moretti, G. S. Schorr, L. Thomas, and P. L. Tyack. 2013. "First Direct Measurements of Behavioural Responses by Cuvier's Beaked Whales to Mid-Frequency Active Sonar." *Biology Letters* 9:20130223.

Doty, M. S., and M. Oguri. 1956. "The Island Mass Effect." *Journal du Conseil-Conseil International pour l'Exploration de la Mer* 22:33–37.

Faerber, M. M., and R. W. Baird. 2010. "Does a Lack of Observed Beaked Whale Strandings in Military Exercise Areas Mean No Impacts Have Occurred? A Comparison of Stranding and Detection Probabilities in the Canary and Hawaiian Islands." *Marine Mammal Science* 26:602–613.

Ferreira, I. M., T. Kasuya, H. Marsh, and P. B. Best. 2013. "False Killer Whales (*Pseudorca crassidens*) from Japan and South Africa: Differences in Growth and Reproduction." *Marine Mammal Science* doi: 10.1111/mms.12021.

Foltz, K., R. W. Baird, G. M. Ylitalo, and B. A. Jensen. 2014. "Cytochrome P4501A1 Expression in Blubber Biopsies of Free-Ranging Hawaiian False Killer Whales (*Pseudorca crassidens*) and Other Odontocetes." *Exotoxicology* doi: 10.1007/s10646–014–1300–0.

Goldbogen, J. A., B. L. Southall, S. L. DeRuiter, J. Calambokidis, A. S. Friedlaender, E. L. Hazen, E. A. Falcone, G. S. Schorr, A. Douglas, D. J. Moretti, C. Kyburg, M. F. McKenna, and P. L. Tyack. 2013. "Blue Whales Respond to Simulated Mid-Frequency Military Sonar." *Proceedings of the Royal Society B* 280:20130657.

Henderson, E. E., R. A. Mazano-Roth, S. W. Martin, and B. Matsuyama. 2015. "Impacts of U.S. Navy Training Events on Beaked Whale Foraging Dives in Hawaiian Waters: Update." Available at http://www.navymarinespeciesmonitoring.us/files/1914/3826/9126/Henderson_et_al_2015_Impacts_on_Beaked_Whale_Dives.pdf. Accessed on December 18, 2015.

Herman, L. M., C. S. Baker, P. H. Forestell, and R. C. Antinoja. 1980. "Right Whale *Balaena glacialis* Sightings Near Hawaii: A Clue to the Wintering Grounds?" *Marine Ecology Progress Series* 2:271–275.

Johnston, D. W., M. E. Chapla, L. E. Williams, and D. K. Mattila. 2007. "Identification of Humpback Whale *Megaptera novaeangliae* Wintering Habitat in the Northwestern Hawaiian Islands Using Spatial Habitat Modeling." *Endangered Species Research* 3:249–257.

Karczmarski, L., B. Wursig, G. Gailey, K. W. Larson, and C. Vanderlip. 2005. "Spinner Dolphins in a Remote Hawaiian Atoll: Social Grouping and Population Structure." *Behavioral Ecology* 16:675–685.

Kennedy, A. S., D. R. Salden, and P. J. Clapham. 2011. "First High- to Low-Latitude Match of an Eastern North Pacific Right Whale (*Eubalaena japonica*)." *Marine Mammal Science* doi: 10.1111/j.1748–7692.2011.00539.x.

Kirch, P.V. 2010. "When Did the Polynesians Settle Hawai'i? A Review of 150 Years of Scholarly Inquiry and a Tentative Answer." Available at http://www.academia.edu/1034367/When_Did_the_Polynesians_Settle_Hawaii. Accessed on September 22, 2015.

Lammers, M. O., P. I. Fisher-Pool, W. W. L. Au, C. G. Meyer, K. B. Wong, and R. E. Brainard. 2011. "Humpback Whale *Megaptera novaeangliae* Song Reveals Wintering Activity in the Northwestern Hawaiian Islands." *Marine Ecology Progress Series* 423:261–268.

Mahaffy, S. D., R. W. Baird, D. J. McSweeney, D. L. Webster, and G. S. Schorr. 2015. "High Site Fidelity, Strong Associations and Long-Term Bonds: Short-Finned Pilot Whales off the Island of Hawai'i." *Marine Mammal Science* doi: 10.1111/mms/12234.

Maldini, D. 2003. "Abundance and Distribution Patterns of Hawaiian Odontocetes: Focus on O'ahu." PhD thesis, University of Hawai'i, Mānoa.

Martien, K. K., R. W. Baird, N. M. Hedrick, A. M. Gorgone, J. L. Thieleking, D. J. McSweeney, K. M. Robertson, and D. L. Webster. 2011. "Population Structure of Island-Associated Dolphins: Evidence from Mitochondrial and Microsatellite Markers for Common Bottlenose

Dolphins (*Tursiops truncatus*) around the Main Hawaiian Islands." *Marine Mammal Science* doi: 10.1111/j.1748-7692.2011.00506.x.

Martien, K. K., S. J. Chivers, R. W. Baird, F. I. Archer, A. M. Gorgone, B. L. Hancock-Hanser, D. Mattila, D. J. McSweeney, E. M. Oleson, C. Palmer, V. L. Pease, K. M. Robertson, G. S. Schorr, M. B. Schultz, D. L. Webster, and B. L. Taylor. 2014. "Nuclear and Mitochondrial Patterns of Population Structure in North Pacific False Killer Whales (*Pseudorca crassidens*)." *Journal of Heredity* doi: 10.1093/jhered/esu029.

Martin, S. W., C. R. Martin, B. M. Matsuyama, and E. E. Henderson. 2015. "Minke Whales (*Balaenoptera acutorostrata*) Respond to Navy Training." *Journal of the Acoustical Society of America* 137:2533–2541.

Mazzuca, L., S. Atkinson, B. Keating, and E. Nitta. 1999. "Cetacean Mass Strandings in the Hawaiian Archipelago, 1957–1998." *Aquatic Mammals* 25:105–114.

McSweeney, D. J., R. W. Baird, and S. D. Mahaffy. 2007. "Site Fidelity, Associations and Movements of Cuvier's (*Ziphius cavirostris*) and Blainville's (*Mesoplodon densirostris*) Beaked Whales off the Island of Hawai'i." *Marine Mammal Science* 23:666–687.

McSweeney, D. J., R. W. Baird, S. D. Mahaffy, D. L. Webster, and G. S. Schorr. 2009. "Site Fidelity and Association Patterns of a Rare Species: Pygmy Killer Whales (*Feresa attenuata*) in the Main Hawaiian Islands. *Marine Mammal Science* 25:557–572.

Migaki, G., T. R. Sawa, and J. P. Dubey. 1990. "Fatal Disseminated Toxoplasmosis in a Spinner Dolphin (*Stenella longirostris*)." *Veterinary Pathology* 27:463–464.

Mizroch, S. A., P. B. Conn, and D. W. Rice. 2015. "The Mysterious Sei Whale: Its Distribution, Movements and Population Decline in the North Pacific Revealed by Whaling Data and Recoveries of Discovery-Type Marks." International Whaling Commission Document SC/66a/IA/14.

Mobley, J. R., L. Mazzuca, A. S. Craig, M. W. Newcomer, and S. S. Spitz. 2001. "Killer Whales (*Orcinus orca*) Sighted West of Ni'ihau, Hawai'i. *Pacific Science* 55:301–303.

Mobley, J. R., S. S. Spitz, K. A. Forney, R. Grotefendt, and P. H. Forestell. 2000. "Distribution and Abundance of Odontocete Species in Hawaiian Waters: Preliminary Results of 1993–98 Aerial Surveys." Southwest Fisheries Science Center Administrative Report LJ-00-14C.

Monnahan, C. C., T. A. Branch, and A. E. Punt. 2015. "Do Ship Strikes Threaten the Recovery of Endangered Eastern North Pacific Blue Whales?" *Marine Mammal Science* 31: 279–297.

Moore, J. E., and J. P. Barlow. 2013. "Declining Abundance of Beaked Whales (Family Ziphiidae) in the California Current Large Marine Ecosystem." *PLOS ONE* 8:e52770.

Nitta, E. T. 1991. "The Marine Mammal Stranding Network for Hawaii: An Overview." NOAA Technical report NMFS 98:55–62.

Nitta, E. T., and J. R. Henderson. 1993. "A Review of Interactions between Hawaii's Fisheries and Protected Species." *Marine Fisheries Review* 55:83–92.

Nogelmeier, P., ed. 2010. "Hawaiian Newspaper Translation Project: Fisheries." University of Hawaii Sea Grant College Program BB-10-04.

Norris, K. S. 1974. *The Porpoise Watcher.* New York: W.W. Norton & Co.

Norris, K. S., B. Würsig, R. S. Wells, and M. Würsig. 1994. *The Hawaiian Spinner Dolphin.* Berkeley: University of California Press.

Oleson, E. M., C. H. Boggs, K. A. Forney, M. B. Hanson, D. R. Kobayashi, B. L. Taylor, P. R. Wade, and G. M. Ylitalo. 2010. "Status Review of Hawaiian Insular False Killer Whales (*Pseudorca crassidens*) under the Endangered Species Act." NOAA Technical Memorandum NMFS-PIFSC-22.

Oleson, E. M., A. Širović, A. R. Bayless, and J. A. Hildebrand. 2014. "Synchronous Seasonal Change in Fin Whale Song in the North Pacific." *PLOS ONE* 9:e115678.

Pack, A. A., L. M. Herman, S. S. Spitz, A. S. Craig, S. Hakala, M. H. Deakos, E. Y. K. Herman, A. J. Milette, E. Carroll, S. Levitt, and C. Lowe. 2012. "Size-Assortative Pairing and Discrimination of Potential Mates by Humpback Whales in the Hawaiian Breeding Grounds." *Animal Behaviour* 84:983–993.

Plön, S. 2004. "The Status and Natural History of Pygmy (*Kogia breviceps*) and Dwarf (*K. sima*) Sperm Whales off Southern Africa." PhD thesis, Rhodes University, Grahamstown, South Africa.

Pryor, K. 1975. *Lads before the Wind:Diary of a Dolphin Trainer.* North Bend, WA: Sunshine Books.

Rankin, S., T. F. Norris, M. A. Smultea, C. Oedekoven, A. M. Zoidis, E. Silva, and J. Rivers. 2007. "A Visual Sighting and Acoustic Detections of Minke Whales, *Balaenoptera acutorostrata* (Cetacea: Balaenopteridae), in Nearshore Hawaiian waters." *Pacific Science* 61:395–398.

Reeves, R. R., S. Leatherwood, and R. W. Baird. 2009. "Evidence of a Possible Decline since 1989 in False Killer Whales (*Pseudorca crassidens*) around the Main Hawaiian Islands." *Pacific Science* 63:253–261.

Rice, D. W. 1960. "Distribution of the Bottle-Nosed Dolphin in the Leeward Hawaiian Islands." *Journal of Mammalogy* 41:407–408.

Richards, L. P. 1952. "Cuvier's Beaked Whale from Hawaii." *Journal of Mammalogy* 33:255.

Rocha, R. C., P. J. Clapham, and Y. V. Ivashchenko. 2014. "Emptying the Oceans: A Summary of Industrial Whaling Catches in the 20th Century." *Marine Fisheries Review* 76:37–48.

Rowntree, V., J. Darling, G. Silber, and M. Ferrari. 1980. "Rare Sightings of a Right Whale (*Eubalaena glacialis*) in Hawaii." *Canadian Journal of Zoology* 58:308–312.

Schorr, G. S., R. W. Baird, M. B. Hanson, D. L. Webster, D. J. McSweeney, and R. D. Andrews. 2009. "Movements of Satellite-Tagged Blainville's Beaked Whales off the Island of Hawai'i." *Endangered Species Research* 10:203–213.

Schorr, G. S., E. A. Falcone, D. J. Moretti, and R. D. Andrews. 2014. "First Long-Term Behavioral Records from Cuvier's Beaked Whales (*Ziphius cavirostris*) Reveal Record-Breaking Dives." *PLOS ONE* 9(3): e92633.

Scott, M. D., and J. G. Cordaro. 1987. "Behavioral Observations of the Dwarf Sperm Whale, *Kogia simus*." *Marine Mammal Science* 3:353–354.

Seki, M. P., R. Lumpkin, and P. Flament. 2002. "Hawaii Cyclonic Eddies and Blue Marlin Catches: The Case Study of the 1995 Hawaiian International Billfish Tournament." *Journal of Oceanography* 58:739–745.

Shallenberger, E. W. 1981. "The Status of Hawaiian Cetaceans." Report No. MMC-77/23 of the U.S. Marine Mammal Commission.

Shane, S. H., L. Teply, and L. Costello. 1993. "Life-Threatening Contact between a Woman and a Pilot Whale Captured on Film." *Marine Mammal Science* 9:331–336.

Shomura, R. S., and T. S. Hida. 1965. "Stomach Contents of a Dolphin Caught in Hawaiian Waters." *Journal of Mammalogy* 46:500–501.

Smultea, M. A., T. A. Jefferson, and A. M. Zoidis. 2010. "Rare Sightings of a Bryde's Whale (*Balaenoptera edeni*) and Sei Whales (*B. borealis*) (Cetacea: Balaenopteridae) Northeast of O'ahu, Hawai'i." *Pacific Science* 64:449–457.

Southall, B. L., R. Braun, F. M. D. Gulland, A. D. Heard, R. W. Baird, S. M. Wilkin, and T. K. Rowles. 2006. "Hawaiian Melon-Headed Whale (*Peponocephala electra*) Mass Stranding Event of July 3–4, 2004." NOAA Technical Memorandum NMFS-OPR-31.

Southall, B. L., T. Rowles, F. Gulland, R. W. Baird, and P. D. Jepson. 2013. "Final Report of the Independent Scientific Review Panel Investigating Potential Contributing Factors to a 2008 Mass Stranding of Melon-Headed Whales (*Peponocephala electra*) in Antsohihy, Madagascar." International Whaling Commission.

Stafford, K. M., S. L. Nieukirk, and C. G. Fox. 2001. "Geographic and Seasonal Variation of Blue Whale Calls in the North Pacific." *Journal of Cetacean Research and Management* 3:65–76.

Thompson, P. O., and W. A. Friedl. 1982. "A Long Term Study of Low Frequency Sounds from Several Species of Whales off Oahu, Hawaii." *Cetology* 45:1–19.

Tomich, P. Q. 1986. *Mammals in Hawai'i.* 2nd ed. Honolulu: Bishop Museum Press.

Tyack, P. L., W. M. X. Zimmer, D. Moretti, B. L. Southall, D. E. Claridge, J. W. Durban, C. W. Clark, A. D'Amico, N. DiMarzio, S. Jarvis, E. McCarthy, R. Morrissey, J. Ward, and I. L. Boyd. 2011. "Beaked Whales Respond to Simulated and Actual Navy Sonar." *PLOS ONE* 6: e17009.

Tyne, J. 2015. "A Scientific Foundation for Informed Management Decisions: Quantifying the Abundance, Important Habitat and Cumulative Exposure of the Hawaii Island Spinner Dolphin (*Stenella longirostris*) Stock to Human Activities." PhD thesis, Murdoch University.

West, K. L., S. Sanchez, D. Rotstein, K. M. Robertson, S. Dennison, G. Levine, N. Davis, D. Schofield, C. W. Potter, and B. Jensen. 2013. "A Longman's Beaked Whale (*Indopacetus pacificus*) Strands in Maui, Hawaii, with First Case of Morbillivirus in the Central Pacific." *Marine Mammal Science* 29:767–776.

West, K. L., W. A. Walker, R. W. Baird, W. White, G. Levine, and E. Brown. 2009. "Diet of Pygmy Sperm Whales (*Kogia breviceps*) in the Hawaiian Archipelago." *Marine Mammal Science* 25:931–943.

Whitehead, H. 2003. *Sperm Whales.* Chicago: University of Chicago Press.

Wilkes, C. 1845. *Narrative of the United States Exploring Expedition during the Years 1838, 1839, 1840, 1841, 1842.* Philadelphia: Lea and Blanchard.

Woodworth, P. A., G. S. Schorr, R. W. Baird, D. L. Webster, D. J. McSweeney, M. B. Hanson, R. D. Andrews, and J. J. Polovina. 2011. "Eddies as Offshore Foraging Grounds for Melon-Headed Whales (*Peponocephala electra*)." *Marine Mammal Science* doi: 10.1111/j.1748-7692.2011.00509.x.

Ylitalo, G. M., R. W. Baird, G. K. Yanagida, D. L. Webster, S. J. Chivers, J. L. Bolton, G. S. Schorr, and D. J. McSweeney. 2009. "High Levels of Persistent Organic Pollutants Measured in Blubber of Island-Associated False Killer Whales (*Pseudorca crassidens*) around the Main Hawaiian Islands." *Marine Pollution Bulletin* 58:1932–1937.

INDEX

Page numbers in boldface type refer to illustrations

INDEX

INDEX

ABOUT THE AUTHOR

ROBIN W. BAIRD is a biologist who has been studying whales and dolphins since 1985. Baird first began working with killer whales in British Columbia, Canada, and in 1994 received a PhD in biological sciences from Simon Fraser University (Burnaby, Canada), with his thesis on foraging behavior and ecology of mammal-eating killer whales. He was one of the founders of the Stranded Whale and Dolphin Program of British Columbia in 1987, and co-coordinated the program until 1995. In the 1990s and early 2000s he was involved in whale and dolphin studies in a number of locations around the world, including Mexico, Japan, New Zealand, Iceland, Ireland, Italy, Nova Scotia, Washington, California, and North Carolina. He moved to Hawai'i in December 1998 and began working with a number of species of whales and dolphins, and while based in Olympia, Washington, since 2003, he has worked in Hawai'i every year since 1999. He is a member of the Committee of Scientific Advisors of the U.S. Marine Mammal Commission, the IUCN Cetacean Specialist Group, and the False Killer Whale Take Reduction Team. He has authored or coauthored ninety-nine publications in peer-reviewed journals. Robin is a research biologist with Cascadia Research Collective, a nonprofit research and education organization based in Olympia, Washington, and can be reached at rwbaird@cascadiaresearch.org.